ATLAS
OF ZEOLITE
FRAMEWORK TYPES

ATLAS OF ZEOLITE FRAMEWORK TYPES

Sixth Revised Edition

dedicated to

Walter M. Meier

*co-author of the first edition of the Atlas and
co-founder of the IZA Structure Commission*

on the occasion of his 80th birthday

Christian Baerlocher and Lynne B. McCusker
Laboratory of Crystallography
ETH Zurich
8093 Zurich, Switzerland

David H. Olson
Department of Chemistry and Chemical Biology
Rutgers University
Piscataway, NJ 08854, USA

Published on behalf of the
Structure Commission of the International Zeolite Association by

ELSEVIER

Amsterdam ● Boston ● Heidelberg ● London ● New York ● Oxford
Paris ● San Diego ● San Francisco ● Singapore ● Sydney ● Tokyo

Elsevier
Radarweg 29, PO Box 211, 1000 AE Amsterdam, The Netherlands
Linacre House, Jordan Hill, Oxford OX2 8DP, UK

First edition 1978, published by the Structure Commission of the International Zeolite Association
Second revised edition, 1987, published by Butterworth-Heinemann
Third revised edition, 1992, published by Butterworth-Heinemann
Fourth revised edition, 1996, published by Elsevier Science
Fifth revised edition, first impression, 2001, published by Elsevier Science
Fifth revised edition, second impression 2001, published by Elsevier Science
Sixth revised edition, first impression 2007

Library of Congress Cataloging-in-Publication Data
A catalog record for this book is available from the Library of Congress

British Library Cataloguing in Publication Data
A catalogue record for this book is available from the British Library

ISBN: 978-0-444-53064-6

For information on all Elsevier publications
visit our website at books.elsevier.com

Printed and bound in the United Kingdom

Transferred to Digital Print 2011

TABLE OF CONTENTS

MEL	ZSM-11		SAS	STA-6
MEP	Melanophlogite		SAT	STA-2
MER	Merlinoite		SAV	Mg-STA-7
MFI	ZSM-5		SBE	UCSB-8Co
MFS	ZSM-57		SBS	UCSB-6GaCo
MON	Montesommaite		SBT	UCSB-10GaZn
MOR	Mordenite		SFE	SSZ-48
MOZ	ZSM-10		SFF	SSZ-44
MSE	MCM-68		SFG	SSZ-58
MSO	MCM-61		SFH	SSZ-53
MTF	MCM-35		SFN	SSZ-59
MTN	ZSM-39		SFO	SSZ-51
MTT	ZSM-23		SGT	Sigma-2
MTW	ZSM-12		SIV	SIZ-7
MWW	MCM-22		SOD	Sodalite
NAB	Nabesite		SOS	SU-16
NAT	Natrolite		SSY	SSZ-60
NES	NU-87		STF	SSZ-35
NON	Nonasil		STI	Stilbite
NPO	Nitridophosphate-1		STT	SSZ-23
NSI	Nu-6(2)		SZR	SUZ-4
OBW	OSB-2		TER	Terranovaite
OFF	Offretite		THO	Thomsonite
OSI	UiO-6		TOL	ordered Tounkite
OSO	OSB-1		TON	Theta-1
OWE	UiO-28		TSC	Tschörtnerite
-PAR	Partheite		TUN	TNU-9
PAU	Paulingite		UEI	Mu-18
PHI	Phillipsite		UFI	UZM-5
PON	IST-1		UOZ	IM-10
RHO	Rho		USI	IM-6
-RON	Roggianite		UTL	IM-12
RRO	RUB-41		VET	VPI-8
RSN	RUB-17		VFI	VPI-5
RTE	RUB-3		VNI	VPI-9
RTH	RUB-13		VSV	VPI-7
RUT	RUB-10		WEI	Weinebeneite
RWR	RUB-24		-WEN	Wenkite
RWY	UCR-20		YUG	Yugawaralite
SAO	STA-1		ZON	ZAPO-M1

Appendices

PREFACE

In this sixth edition of the *Atlas of Zeolite Framework Types*, data are presented for each of the 176 unique zeolite Framework Types that had been approved and assigned a 3-letter code by the Structure Commission of the IZA (IZA-SC) by February 2007. Six years ago, this number was 133. The number of new verified Framework Types reflects the vibrant activity that persists within the zeolite community. In 1970, the paper that can be considered to be the forerunner of the first edition of the *Atlas* was published[1] and it described the 27 zeolite framework structures known at the time. With each edition of the *Atlas*, this number has grown and the exponential trend continues. In that initial paper, the foundation for a systematic description of zeolite framework structures was laid, and we are still using these basic concepts. It is with pleasure and gratitude for this pioneering work that we dedicate this edition of the *Atlas* to Walter M. Meier, co-author of that first *Atlas* and co-founder of the Structure Commission of the IZA, on the occasion of his 80th birthday.

Each of the zeolite Framework Types included in this book has been examined by the members of the IZA-SC to verify that it conforms to the IZA-SC definition of a zeolite, that it is unique, and that the structure has been satisfactorily proven. The three-letter code that is then assigned is officially recognized by IUPAC, and is used in the nomenclature recommended by IUPAC for these materials.[2] The rules used by the IZA-SC are given in Appendix B.

As a frequently quoted work of reference, the *Atlas* must be updated on a regular basis to be of full use. Not only must new Framework Types be added, but corrections and new information on existing entries must also be disseminated. To do this, the IZA-SC maintains a freely available searchable database on the internet at http://www.iza-structure.org/databases/. Although all the information in this book (and more) is available at that website, even the most technologically-minded individuals will admit that it is sometimes easier to access the desired information in a book. This is why we continue to publish a hard copy of the most used data on a periodic basis.

In this edition we have prepared new stereo drawings of all Framework Types, added over 250 references, and for the first time, included some composite building units that are common to more than one Framework Type. To make room for the latter, we have removed the loop configurations (still available on the web), reasoning that the vertex symbols provide very similar information. To minimize errors, the data sheets have all been generated directly from the zeolite database that is used for the online version of the *Atlas*.

We wish to acknowledge the assistance and collaboration of many fellow scientists, who have provided us with information and suggested improvements. We are particularly indebted to Henk van Koningsveld, who determined the secondary building units for all the new frameworks, and the members of the IZA Structure Commission, who provided additional information and proofread tirelessly. It does not seem to be possible to assemble such a compilation without mistakes or oversights, so we will be grateful for any corrections and/or additions that you communicate to us.

February 2007   Christian Baerlocher   Lynne B. McCusker   David H. Olson

2

Structure Commission of the International Zeolite Association

Previous IZA Special Publications:

W.M. Meier and D.H. Olson, *Atlas of Zeolite Structure Types* (1978)

W.M. Meier and D.H. Olson, *Atlas of Zeolite Structure Types*, 2nd ed. (1987)

W.M. Meier and D.H. Olson, *Atlas of Zeolite Structure Types*, 3rd ed. (1992)

W.M. Meier, D.H. Olson and Ch. Baerlocher, *Atlas of Zeolite Structure Types*, 4th ed. (1996)

Ch. Baerlocher, W.M. Meier and D.H. Olson, *Atlas of Zeolite Framework Types*, 5th ed. (2001)

W.J. Mortier, *Compilation of Extra Framework Sites in Zeolites* (1982)

R. von Ballmoos, *Collection of Simulated XRD Powder Patterns for Zeolites* (1984)

R. von Ballmoos and J.B. Higgins, *Collection of Simulated XRD Powder Patterns for Zeolites*, 2nd ed. (1990)

M.M.J. Treacy, J.B. Higgins and R. von Ballmoos, *Collection of Simulated XRD Powder Patterns for Zeolites*, 3rd ed. (1996)

M.M.J. Treacy and J.B. Higgins, *Collection of Simulated XRD Powder Patterns for Zeolites*, 4th ed. (2001)

H. Robson and K.P. Lillerud, *Verified Synthesis of Zeolitic Materials* (1998)

INTRODUCTION AND EXPLANATORY NOTES

Zeolites and zeolite-like materials do not comprise an easily definable family of crystalline solids. A simple criterion for distinguishing zeolites and zeolite-like materials from denser tectosilicates is based on the framework density (FD), the number of tetrahedrally coordinated framework atoms (T-atoms) per 1000 $Å^3$. Figure 1 shows the distribution of these values for porous and dense frameworks, whose structures are well established[3]. A gap is clearly recognizable between zeolite-type and dense tetrahedral framework structures. The maximum FD for a zeolite ranges from 19 to over 21 T-atoms per 1000 $Å^3$, depending on the type of smallest ring present, whereas the minimum for denser structures ranges from 20 to 22. Strictly speaking the boundaries defined in Figure 1 for the framework densities only apply to fully crosslinked frameworks, so interrupted frameworks have not been included.

For each Framework Type Code (see below), two pages of data are included in this *Atlas*. The first page lists the information that characterizes the Framework Type. This includes crystallographic data (highest possible space group, cell constants of the idealized framework), coordination sequences, vertex symbols and composite building units. Taken together, the coordination sequences and the vertex symbols define the Framework Type. On the second page, data for the Type *Material* (i.e. the real material on which the idealized Framework Type is based) can be found. Although the channel dimensionality is a property of the Framework Type, the channel description also includes the observed ring dimensions, and must therefore refer to the Type Material. For each Framework Type, a list of isotypic materials with the corresponding references is also given. The different entries in the data sheets are described in more detail below.

Framework Type Page

Framework Type Code

Following the rules set up by an IUPAC Commission on Zeolite Nomenclature in 1979[4], designations consisting of three capital letters (in boldface type) have been used throughout. The codes are generally derived from the names of the Type Materials (see Appendix G) and do not include numbers and characters other than capital Roman letters. The assignment of Framework Type Codes is subject to review and clearance by the IZA Structure Commission according to a decision of the IZA Council (taken at the time of the 7th IZC in Tokyo, 1986). Codes are only assigned to established structures that satisfy the rules of the IZA Structure Commission (see Appendix B for a listing of these rules). For interrupted frameworks, the 3-letter code is preceded by a hyphen. These mnemonic codes should not be confused or equated with actual materials. They only describe and define the network of the corner sharing tetrahedrally coordinated framework atoms. Thus, designations such as NaFAU are untenable. However, a material can be described using the IUPAC crystal chemical formula[2], as |Na$_{58}$| [Al$_{58}$Si$_{134}$ O$_{384}$]-**FAU** or |Na-| [Al-Si-O]-**FAU**. Note that the chemical elements must be enclosed within the appropriate (boldface) brackets (i.e. | | for guest species and [ ] for the framework host)

4

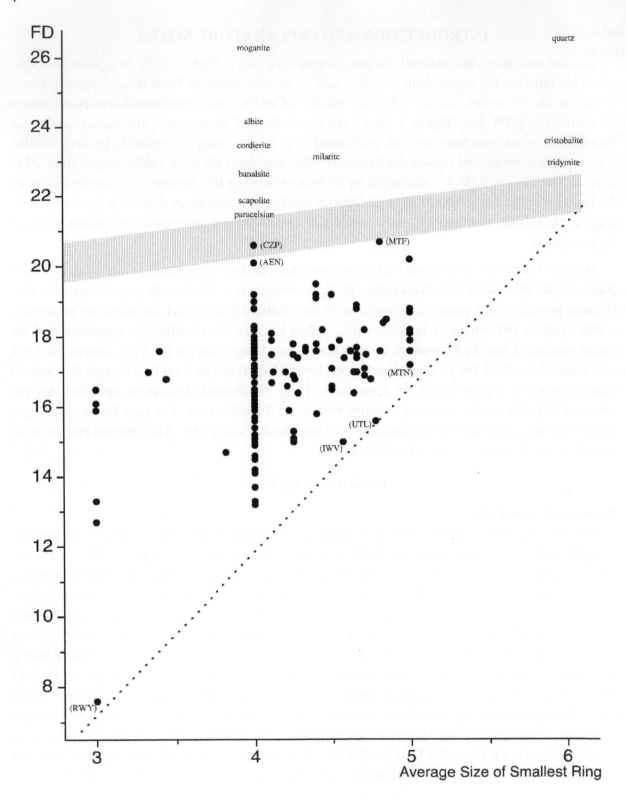

Fig. 1. Framework density calculated for the idealized SiO_2 framework vs. average size of the smallest ring in the structure.

and that the stoichiometry may, but does not have to be specified. Framework Types do not depend on composition, distribution of the T-atoms (Si, Al, P, Ga, Ge, B, Be, etc.), cell dimensions or symmetry.

The Framework Types have been arranged in alphabetical order according to the Framework Type Code, because structural criteria alone do not provide an unambiguous classification scheme. This also facilitates later insertion of new codes and allows simple indexing. The Framework Type Code is given at the top of each page. On the first page this is supplemented with the maximum space group symmetry for the framework, and on the second page with the full name of the Type Material. In this edition, the new designation "e" in space groups with double-glide planes has been used (e.g. *Cmma* is now *Cmme*).

Framework drawing

A stereographic drawing of the framework, generated using the program CrystalMaker[5] is presented for each Framework Type. Although the depth fading helps in viewing the drawings, the use of a stereo viewer is recommended (can be obtained from any electron microscopy supply house). The coordinates of the idealized, highest symmetry structures have been used for these drawings. Only the positions of the T-atoms are shown. The T-O-T bridges are represented by straight lines. This idealization makes it easier to visualize the topology and the basic features of these (often complex) framework structures. The unit cell is outlined and the orientation of the axes indicated. For most frameworks, more than a unit cell is shown, but for cases where the figures would have been too complex, only a part of the unit cell is depicted. In these cases a projection along a major axis is shown to help in visualizing the build-up of the complete framework.

Idealized cell parameters

The idealized cell parameters were obtained from a DLS-refinement[6] using the given (highest possible) symmetry of the Framework Type. The refinement was carried out assuming a (sometimes hypothetical) SiO_2 composition and with the following prescribed interatomic distances: $d_{Si-O} = 1.61Å$, $d_{O-O} = 2.629Å$ and $D_{Si-Si} = 3.07Å$ using the weights of 2.0, 0.61 and 0.23, respectively.

Coordination sequences (CS) and vertex symbols

The concept of coordination sequences was originally introduced by Brunner and Laves[7] and first applied to zeolite frameworks by Meier and Moeck[8]. In a typical zeolite framework, each T-atom is connected to $N_1 = 4$ neighboring T-atoms through oxygen bridges. These neighboring T-atoms are then linked in the same manner to N_2 T-atoms in the next shell. The latter are connected to N_3 T-atoms etc. Each T-atom is counted only once. In this way, a coordination sequence can be determined for each T-atom of the 4-connected net of T-atoms. It follows that

$$N_0 = 1 \qquad N_1 \leq 4 \qquad N_2 \leq 12 \qquad N_3 \leq 36... \qquad N_k \leq 4 \bullet 3^{k-1}$$

CS's are listed from N_1 up to N_{10} for each topologically distinct T-atom in the framework structure along with the site multiplicity and the site symmetry (both in parenthesis).

The vertex symbol was first used in connection with zeolite-type networks by M. O'Keefe and S.T. Hyde[9]. This symbol indicates the size of the smallest ring associated with each of the 6 angles of a tetrahedron (T-atom). The symbols for opposite pairs of angles are grouped together. For **FAU** the vertex symbol reads 4· 4· 4· 6· 6· 12, indicating that one pair of opposing angles contains 4-rings, a second pair a 4-ring and a 6-ring, and the final pair a 6-ring and a 12-ring. It is useful for determining the smallest rings in a framework. In the case of **DOH**, for example, the vertex symbols for the four T-atoms are 5· 5· 5· 5· 5· 6, 4· 5· 5· 6· 5· 6, 5· 5· 5· 5· 5· 6 and 5· 5· 5· 5· 5· 5, so the smallest rings are 4- and 5-rings, (4.75 in Figure 1). Sometimes more than one ring of the same size is found for a single angle. This is indicated with a subscript like 6_2 or 8_2. An asterisk in the vertex symbol indicates that no ring is formed for that angle.

The coordination sequence and the vertex symbol together appear to be unique for a particular framework topology. That is, they can be used to distinguish different zeolite Framework Types unambiguously. In this way, isotypic frameworks can be recognized easily.

Secondary building units (SBU's)

Zeolite frameworks can be thought to consist of finite or infinite (i.e. chain- or layer-like) component units. The finite units which have been found to occur in at least two tetrahedral frameworks are shown in Appendix C. These secondary building units (primary building units are TO_4 tetrahedra), which contain up to 16 T-atoms, are derived assuming that the entire framework is made up of one type of SBU only. It should be noted that SBU's are invariably non-chiral (neither left- nor right-handed) and a unit cell always contains an integral number of them. As far as practicable, all possible SBU's have been listed. The number of observed SBU's has increased from 20 in 2001 to 23. For Framework Types with an SBU that only occurs once or for which combinations of SBU's are necessary, the reader is referred to the *Compendium of Zeolite Framework Types. Building Schemes and Type Characteristics* by H. van Koningsveld[10], where all SBU's are listed in detail. The symbols given below the drawings in Appendix C are used in the data sheets. If more than one SBU is possible for a given Framework Type, all are listed. The number given in parenthesis in Appendix C indicates the frequency of the occurrence of that SBU. The SBU's are only theoretical topological building units and should not be considered to be or equated with species that may be in the solution/gel during the crystallization of a zeolitic material.

Framework description

For all 19 Framework Types belonging to the so-called ABC-6-family, the ABC stacking sequence is given. Some other structural relationships which are thought to be helpful are also listed.

Composite Building Units (CBU's)

Some units (e.g. double 6-ring, cancrinite cage, alpha cavity, double crankshaft chain) appear in several different framework structures, and can be useful in identifying relationships between Framework Types. Smith has compiled an exhaustive list of such units, not only for zeolite structures but also for hypothetical 3-dimensional 4-connected nets[11]. In his *Compendium*[10], van Koningsveld has also included an extensive list of them. Here we have arbitrarily selected just 47 Composite Building Units and five chains (Appendix D) that are found in at least two different Framework Types. These are different from secondary building units in that they are not required to be achiral, and cannot necessarily be used to build the entire framework. To facilitate communication, each unit has been assigned a lower case italic three-character designation. With the exception of the double 4-, 6- and 8-rings (*d4r*, *d6r* and *d8r*, respectively), a code corresponding to one of the Framework Types containing the CBU has been used for this purpose. A comparison of the notation used by Smith, van Koningsveld and here is also given in Appendix D.

Materials with this Framework Type

Under this heading, as-synthesized materials that have the same Framework Type but different chemical composition or have a different laboratory designation are listed. Materials obtained by post-synthesis treatment (e.g. ion exchange, dealumination, etc.) are generally not included. The Type Material (defined on the second page) is given first and marked with an asterisk. These materials are also listed in the Isotypic Material Index at the end of the book.

References

The list of references cited is far from complete. As a general rule, references for the Type Material are to the work that first established the Framework Type and to subsequent work adding significant information regarding the framework topology. Thus papers on non-framework species have not been included. References for other materials with a specific framework type are limited to work in which sufficient data are provided to establish the framework type.

For most of the codes from **ABW** to **RHO**, complete references, cell constant data, space groups, site symmetries, symmetry relationships, structural diagrams, positional coordinates and chemical compositions for all crystal structure determinations published up to April 2000 can be found in the Landolt-Börnstein Series on *Microporous and other framework materials with zeolite-type structures* edited by W.H. Baur and R.X. Fischer[12].

Type Material Page

The Type Material is the species first used to establish the Framework Type. Detailed information about the material is given on this page.

Crystal chemical data

The chemical composition, expressed in terms of unit cell contents, has been idealized where necessary for simplicity. The chemical formula is given according to IUPAC recommendations[2]. For each Type Material, the space group and cell parameters listed are taken from the reference cited. In many instances, further refinement of the structure taking into account ordering etc. would yield a lower symmetry. It should also be noted that the space group and other crystallographic data related to the Type Material structure do not necessarily apply to isotypes.

In some cases, the space group setting of the Type Material differs from that of the Framework Type. In these cases, the relationship between the two unit cells is given. This relationship is important when comparing the orientation of the channel direction and the viewing direction of ring drawings (both are given for the axis orientation of the Type Material) with that of the framework drawing.

Framework density (FD)

The framework density is defined as the number of T-atoms per 1000 Å^3. The number given refers to the Type Material. For non-zeolitic framework structures, these values tend to be at least 20 to 21 T/1000 Å^3, while for zeolites with fully crosslinked frameworks, the observed values range from ~12.1, for structures with the largest pore volume, to ~20.6. To date, FD's of less than 12 have only been encountered for the interrupted framework cloverite (-**CLO**)[13], for the sulfide UCR-20[14] (**RWY**), and for hypothetical networks[15]. The FD is obviously related to the pore volume but does not reflect the size of the pore openings. For some of the more flexible zeolite structures, the FD values can vary appreciably. In these cases (e.g. gismondine) values are given for the Type Material and for the framework in its most expanded state. The flexibility of a framework structure is, to some extent, revealed by the possible variation in FD. FD values may also depend on chemical composition.

Channels

A shorthand notation has been adopted for the description of the channels in the various frameworks. Each system of equivalent channels has been characterized by

- the channel direction (relative to the axes of the Type Material structure),
- the number of T-atoms (in bold type) forming the rings controlling diffusion through the channels, and
- the crystallographic free diameters of the channels in Angstrom units.

The number of asterisks in the notation indicates whether the channel system is one-, two- or three-dimensional. In most cases, the smaller openings simply form windows (rather than channels) connecting larger cavities. Interconnecting channel systems are separated by a double arrow (↔). A vertical bar (|) means that there is no direct access from one channel system to the other. The examples below have been selected to illustrate the use of this notation.

Cancrinite	[001] **12** 5.9 x 5.9*
Offretite	[001] **12** 6.7 x 6.8* ↔ ⊥ [001] **8** 3.6 x 4.9**
Mordenite	[001] **12** 6.5 x 7.0* ↔ {[010] **8** 3.4 x 4.8 ↔ [001] **8** 2.6 x 5.7}*
Zeolite Rho	<100> **8** 3.6 x 3.6*** | <100> **8** 3.6 x 3.6***
Gismondine	{[100] **8** 3.1 x 4.5 ↔ [010] **8** 2.8 x 4.8}***

Cancrinite is characterized by a 1-dimensional system of channels parallel to [001] (or c) with circular 12-ring apertures. In offretite, the main channels are similar but they are interconnected at right angles by a 2-dimensional system of 8-ring channels, and thus form a 3-dimensional channel system. The channel system in mordenite is essentially 2-dimensional with somewhat elliptical 12-ring apertures. The 8-ring limiting diffusion in the [001] direction is an example of a highly puckered aperture. Zeolite rho is an example of a Framework Type containing two non-interconnecting 3-dimensional channel systems which are displaced with respect to one another (<100> means there are channels parallel to all crystallographically equivalent axes of the cubic structure, i.e., along [100], [010] and [001]). In gismondine, the channels parallel to [100] together with those parallel to [010] give rise to a 3-dimensional channel system which can be pictured as an array of partially overlapping tubes.

A summary of the channel systems, ordered by decreasing number of T-atoms in the largest ring, is given in Appendix E. The free diameter values (effective pore width) given in the channel description and on the ring drawings are based upon the atomic coordinates of the *Type Material* and an oxygen radius of 1.35Å. Both minimum and maximum values are given for non-circular apertures. In some instances, the corresponding interatomic distance vectors are only approximately coplanar, in other cases the plane of the ring is not normal to the direction of the channel. Close inspection of the framework and ring drawings should provide qualitative evidence of these factors. Some ring openings are defined by a very complex arrangement of oxygen atoms, so in these cases other short interatomic distances that are not listed may also be observed. It should be noted that crystallographic free diameters may depend upon the hydration state of the zeolite, particularly for the more flexible frameworks. It should also be borne in mind that effective free diameters can be affected by non-framework cations and may also be temperature dependent.

Note that the channel direction is given for the axis orientation of the *Type Material*. This orientation may be different from the orientation given in the framework drawing (see the cell relationship given under "crystal chemical data" for these cases).

Stability

In some cases, the Type Material is not stable to heating and/or removal of the template. This has been indicated, if the information was available.

Ring drawings

Stereographic drawings of the limiting channel windows are presented for all Framework Types. In contrast to the framework drawings, all atoms are shown in the ring drawings. Their positions are based on the crystal structure of the Type Material, and therefore the ring dimensions and the viewing direction are also those of the Type Material. As explained in the crystal chemical data section, for a few Type Materials, the orientation of the crystallographic axes is different from that given for the Framework Type. In these cases, the relationship given in the "crystal chemical data" section must be applied when comparing the viewing direction of the ring drawings with that of the framework drawing.

Supplementary Information

Topological densities

The coordination sequences (CS) can be used to calculate a topological density (TD). As might be expected, the CS is a periodic function. This has been established for all observed framework topologies by Grosse–Kunstleve, Brunner and Sloane[16]. They showed that the CS of any T–atom can be described exactly by a set of p quadratic equations

$$N_k = a_i k^2 + b_i k + c_i \qquad \text{for } k = i + np, \qquad n = 0,1,2,\ldots \text{and} \quad i = 1,2,3, \ldots p$$

For example, the CS of **ABW** ($N_1= 4$ $N_2= 10$ $N_3= 21$ $N_4=36$ $N_5= 54$ $N_6= 78$ $N_7= 106$, etc.) is exactly described by a set of three quadratic equations (p=3), namely

$$N_k = 19/9\, k^2 + 1/9\, k + 16/9 \qquad \text{for } k = 1 + 3n, \qquad n = 0,1,2,\ldots$$

$$N_k = 19/9\, k^2 - 1/9\, k + 16/9 \qquad \text{for } k = 2 + 3n, \qquad n = 0,1,2,\ldots$$

$$N_k = 19/9\, k^2 - 0\, k + 2 \qquad \text{for } k = 3 + 3n, \qquad n = 0,1,2,\ldots$$

The number of equations p necessary to calculate all members of a particular coordination sequence varies from p=1 for **SOD** and p=42 for **FAU** to p=140,900,760 for **EUO**.

With growing index k (the shell number of the CS), the linear and constant coefficients, b_i and c_i, respectively, become less and less important. Therefore we can define the exact topological density TD as the mean of all a_i divided by the dimensionality of the topology (i.e. 3 for zeolites)

$$TD = \frac{<a_i>}{3} = \frac{1}{3p} \sum_{i=1}^{p} a_i$$

This TD is the same for all T atoms in a given structure. For some frameworks, this calculation can take quite a long time, so an approximation valid to ±0.001 has been used to calculate the values given in Appendix F for each of the Framework Types. The value for $<a_i>$ has been approximated as the mean of a_i for the last 100 terms of a CS with 1000 terms (TD1000:100), weighted with the multiplicity of the atom position, and divided by three (dimensionality). The value for TD_{10} (sum of the CS values for N_0 to N_{10}), which was listed in the first four editions of the Atlas, is also given for comparison. There is a simple relationship between TD and TD_{10}: $TD_{10} \sim TD *1155$. Since TD_{10} is an approximation, i.e. it is 'arbitrarily' terminated at N_{10}, the values obtained by this formula deviate by 11% for **–CLO** and 5% for **FAU** but the differences are generally below 3%. It seems that for very open structures, 10 steps are not sufficient for a satifactory convergence. The correlation factor between the exact topological density TD and the framework density FD is 0.82.

Origin of 3-letter codes and Type Material names

The derivation of the 3-letter codes for the zeolite minerals is fairly obvious, because the code generally consists of the first 3 letters of the mineral name. For the synthetic materials this is sometimes more obscure. One reason for this is that numbers are frequently included to distinguish different products from a particular lab, and these numbers cannot be transferred directly to the framework code. To help the reader better understand the origin of the codes, a table that includes all Framework Type codes derived from synthetic Type Materials is given in Appendix G. In this table, the letters taken for the code are written in bold. Also, an attempt has been made to decipher the origin of the mnemonic sometimes used in the designations of these materials.

Isotypic material index

All materials are listed in alphabetical order in this index. To make the index as informative as possible, all reported materials and designations have been included in this section, provided the Framework Type assignment appears to be reasonably well established. Even a number of occasionally used, but discredited, names of mineral species have been included for the same reason. A full list of obsolete and discredited zeolite mineral names can be found in a report of the subcommittee on zeolites of the International Mineralogical Association[17]. Moreover, the inclusion of a synthetic material's designation in this index must not be interpreted to mean that the designation has been formally recognized or generally accepted. References are to be found on the respective Framework Type data sheets.

References

(1) W.M. Meier and D.H. Olson, *Adv. Chem. Ser.* **101**, 155 (1970)
(2) L.B. McCusker, F. Liebau and G. Engelhardt, *Pure Appl. Chem.* **73**, 381 (2001)
(3) G.O. Brunner and W.M. Meier, *Nature* **337**, 146 (1989)
(4) R.M. Barrer, *Pure Appl. Chem.* **51**, 1091 (1979)
(5) CrystalMaker, a Crystal Structure Program for MacOS Computers. CrystalMaker Software, P.O. Box 183, Bicester, Oxfordshire, OX6 7BS, UK (http://www.crystalmaker.co.uk)
(6) Ch. Baerlocher, A. Hepp. and W.M Meier, "DLS-76, a program for the simulation of crystal structures by geometric refinement". (1978). Lab. f. Kristallographie, ETH, Zürich.
(7) G.O. Brunner and F. Laves, *Wiss. Z. Techn. Univers. Dresden* **20**, 387 (1971) H.2
(8) W.M,. Meier and H.J. Moeck, *J. Solid State Chem.* **27**, 349 (1979)
(9) M. O'Keefe and S.T. Hyde, *Zeolites* **19**, 370 (997)
(10) H. van Koningsveld, *Compendium of Zeolite Framework Types. Building Schemes and Type Characteristics*, Elsevier, Amsterdam, 2007
(11) J.V. Smith, *Tetrahedral frameworks of zeolites, clathrates and related materials*. Subvolume A in Landolt-Börnstein, Numerical Data and Functional Relationships in Science and Technology, New Series, Group IV: Physical Chemistry, Volume 14, Microporous and other Framework Materials with Zeolite-Type Structures, eds. W.H. Baur, R.X. Fischer, Springer, Berlin, 2000
(12) W.H. Baur and R.X. Fischer, *Zeolite Structure Codes ABW to RHO*. Subvolumes B-D in Landolt-Börnstein, Numerical Data and Functional Relationships in Science and Technology, New Series, Group IV: Physical Chemistry, Volume 14, Microporous and other Framework Materials with Zeolite-Type Structures, eds. W.H. Baur, R.X. Fischer, Springer, Berlin, 2000
(13) M. Estermann, L.B. McCusker, Ch. Baerlocher, A. Merrouche and H. Kessler, *Nature* **352**, 320 (1991)
(14) N. Zheng, X. Bu, B. Wang and P. Feng, *Science* **298**, 2366 (2002)
(15) W.M. Meier, *Proc. 7th IZC Tokyo*, Kodansha-Elsevier, 1986, p. 13
(16) R.W. Grosse-Kunstleve, G.O. Brunner and N.J.A. Sloane, *Acta Crystallogr.* **A52**, 879 (1996)
(17) D.S. Coombs, A. Alberti, T. Armbruster, G. Artioli, C. Colella, E. Galli, J.D. Grice, F. Liebau, J.A. Mandarino, H. Minato, E.H. Nickel, E. Passaglia, D.R. Peacor, S. Quartieri, R. Rinaldi, M. Ross, R.A. Sheppard, E. Tillmanns, G. Vezzalini, *Can. Mineral.* **35**, 1571 (1997) or *Mineral. Mag.* **64**, 533 (1998) or *Eur. J. Mineral.* **10**, 1037 (1998)

framework code. To help the reader better understand the origin of the codes, a table that includes all Framework Type codes derived from synthetic Type Materials is given in Appendix C. In this table, the letter taken for the code are written in bold. Also, an attempt has been made to decipher the origin of the mnemonic sonorities used in the designations of these materials.

Isotypic material index

All materials are listed in alphabetical order in this index. To make the index as informative as possible, all reported materials and designations have been included in this section, provided the Framework Type assignment appears to be reasonably well established. Even a number of occasionally used, but discredited, names of critical species have been included for the same reason. A full list of obsolete and discredited zeolite mineral names can be found in a report of the subcommittee on zeolites of the International Mineralogical Association.* Moreover, the inclusion of a synthetic material's designation in this index must not be interpreted to mean that the designation has been formally recognized or generally accepted. References are to be found on the respective Framework Type data sheets.

References

[1] W.M. Meier and D.H. Olson, Adv. Chem. Ser. **101**, 155 (1971).
[2] L.B. McCusker, F. Liebau and G. Engelhardt, Pure Appl. Chem. **73**, 381 (2001).
[3] G.O. Brunner and W.M. Meier, Nature **337**, 146 (1989).
[4] R.M. Barrer, Pure Appl. Chem. **51**, 1091 (1979).
[5] CrystalMaker, a Crystal Structure Program for Mac OS Computers, CrystalMaker Software Ltd., P.O. Box 183, Bicester, Oxfordshire, OX6 7BS, UK (http://www.crystalmaker.co.uk).
[6] Ch. Baerlocher, A. Hepp and W.M. Meier, DLS 76, a program for the simulation of crystal structures by geometric refinement, (1978), Lab. f. Kristallographie ETH Zürich.
[7] G.O. Brunner and L.B. McCusker, XRS, Zeolites **20**, 349 (1971) HP.
[8] W.M. Meier and H.J. Moeck, J. Solid State Chem. **27**, 349 (1979).
[9] M. O'Keefe and S.T. Hyde, Zeolites **19**, 370 (1997).
[10] Ch. Baerlocher, Compendium of Zeolite Framework Types: Building Schemes and Type Characteristics, Elsevier, Amsterdam, 2007.
[11] W.M. Meier, Database of Zeolite Structures and www.iza-structure.org; in Introduction to Zeolite Science and Practice, Studies in Surface Science and Catalysis Volume 137, H. van Bekkum, E.M. Flanigen, P.A. Jacobs and J.C. Jansen, eds. (Elsevier, Amsterdam, 2001).
[12] Database of Zeolite Structures: Volumes in Microporous and other Framework Materials with Porous Structures, (IZA Structure Commission); www.iza-structure.org.
[13] F.H. Baur and R.X. Fischer, Zeolite Structure Codes ABW to BRO, in Landolt-Börnstein, Numerical Data and Functional Relationships in Science and Technology, New Series Group IV Physical Chemistry Volume 14, Microporous and other Framework Materials with Zeolite-Type Structures, eds. W.H. Baur, R.X. Fischer, Springer, Berlin, 2000.
[14] M. Eddaoudi, J. Kim, N. Mc, Ossaher, A.A. Matzger and O.M. Yaghi, Acc. Chem. Res. **34**, 319 (2001).
[15] W.M. Meier, Pure & Appl. Chem., 1986, p. 135.
[16] R.W. Grosse-Kunstleve, G.O. Brunner and N.J.A. Sloane, Acta Cryst. A **52**, 879 (1996).
[17] D.S. Coombs, A. Alberti, T. Armbruster, G. Artioli, C. Colella, E. Galli, J.D. Grice, F. Liebau, J.A. Mandarino, H. Minato, E.H. Nickel, E. Passaglia, D.R. Peacor, S. Quartieri, R. Rinaldi, M. Ross, R.A. Sheppard, E. Tillmanns and G. Vezzalini, Can. Mineral. **35**, 1571 (1997), or Eur. J. Mineral. **10**, 1037 (1998).

FRAMEWORK TYPE DATA SHEETS

(arranged in alphabethical order by 3-letter code)

ABW

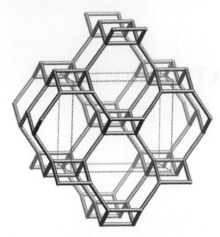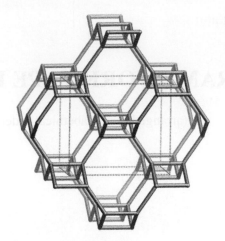

framework viewed along [010]

Idealized cell data: orthorhombic, *Imma*, $a = 9.9$Å, $b = 5.3$Å, $c = 8.8$Å

Coordination sequences and vertex symbols:

T_1 (8,*m*)	4	10	21	36	54	78	106	136	173	214	$4 \cdot 6 \cdot 4 \cdot 6 \cdot 6 \cdot 8_2$

Secondary building units: 8 or 4

Composite building units:

dzc   *abw*

double zigzag chain

Materials with this framework type:

*Li-A (Barrer and White)[1-3]
[Be-As-O]-**ABW**[4,5]
[Be-P-O]-**ABW**[4,6,7]
[Ga-Si-O]-**ABW**[8]
[Zn-As-O]-**ABW**[4,9]
[Zn-P-O]-**ABW**[4]
|(NH₄)-|[Co-P-O]-**ABW**[10]
|(NH₄)-|[Zn-As-O]-**ABW**[11]

|(NH₄)-|[Zn-P-O]-**ABW**[12]
|Cs-|[Mg-P-O]-**ABW**[13]
|Cs-|[Al-Si-O]-**ABW**[14,15]
|Cs-|[Al-Ti-O]-**ABW**[16]
|Li-|[Zn-As-O]-**ABW**[17]
|Li-|[Al-Si-O]-**ABW**[18]
|Li-|[Zn-P-O]-**ABW**[19]
|Li-|[Al-Ge-O]-**ABW**[20]

|Na-|[Zn-P-O]-**ABW**[21]
|Na-|[Co-P-O]-**ABW**[22]
|Rb-|[Cu-P-O]-**ABW**[23]
|Rb-|[Ni-P-O]-**ABW**[24]
|Rb-|[Co-P-O]-**ABW**[13]
|Rb-|[Al-Si-O]-**ABW**[14,15]
|Tl-|[Al-Si-O]-**ABW**[25]
UCSB-3 [26]

Li-A (Barrer and White)   Type Material Data   **ABW**

Crystal chemical data: $|Li_4 (H_2O)_4| [Al_4Si_4O_{16}]$-**ABW**
orthorhombic, $Pna2_1$, $a = 10.31$Å, $b = 8.18$Å, $c = 5.00$Å [2]
(Relationship to unit cell of Framework Type: $a' = a$, $b' = c$, $c' = b$)

Framework density: 19 T/1000Å^3

Channels: [001] **8** 3.4 x 3.8*

8-ring viewed along [001]

References:
(1)   Barrer, R.M. and White, E.A.D. *J. Chem. Soc.*, 1267–278 (1951)
(2)   Kerr, I.S. *Z. Kristallogr.*, **139**, 186–195 (1974)
(3)   Krogh Andersen, E. and Ploug-Sørensen, G. *Z. Kristallogr.*, **176**, 67–73 (1986)
(4)   Gier, T.E. and Stucky, G.D. *Nature*, **349**, 508–510 (1991)
(5)   Harrison, W.T.A., Gier, T.E. and Stucky, G.D. *Acta Crystallogr.*, **C51**, 181–183 (1995)
(6)   Robl, C. and Gobner, V. *J. Chem. Soc., Dalton Trans.*, 1911–1912 (1993)
(7)   Zhang, H., Chen, M., Shi, Z., Bu, X., Zhou, Y., Xu, X. and Zhao, D. *Chem. Mater.*, **13**, 2042–2048 (2001)
(8)   Newsam, J.M. *J. Phys. Chem.*, **92**, 445–452 (1988)
(9)   Feng, P., Zhang, T. and Bu, X. *J. Am. Chem. Soc.*, **123**, 8608–8609 (2001)
(10)   Feng, P., Bu, X., Tolbert, S.H. and Stucky, G.D. *J. Am. Chem. Soc.*, **119**, 2497–2504 (1997)
(11)   Johnson, C.D., Macphee, D.E. and Feldmann, J. *Inorg. Chem.*, **41**, 3588–3589 (2002)
(12)   Bu, X., Feng, P., Gier, T.E. and Stucky, G.D. *Zeolites*, **19**, 200–208 (1997)
(13)   Rakotomahanina Ralaisoa, E.L. *Ph.D. Thesis, Univ. Grenoble* (1972)
(14)   Klaska, R. and Jarchow, O. *Naturwiss.*, **60**, 299 (1973)
(15)   Klaska, R. and Jarchow, O. *Z. Kristallogr.*, **142**, 225–238 (1975)
(16)   Gatehouse, B.M. *Acta Crystallogr.*, **C45**, 1674–1677 (1989)
(17)   Jensen, T.R., Norby, P., Christensen, A.N. and Hanson, J.C. *Microporous Mesoporous Mat.*, **26**, 77–87 (1998)
(18)   Ghobarkar, H. *Cryst. Res. Technol.*, **27**, 1071–1075 (1992)
(19)   Harrison, W.T.A., Gier, T.E., Nicol, J.M. and Stucky, G.D. *J. Solid State Chem.*, **114**, 249–257 (1995)
(20)   Tripathi, A., Kim, S.J., Johnson, G.M. and Parise, J.B. *Microporous Mesoporous Mat.*, **34**, 273–279 (2000)
(21)   Ng, H.Y. and Harrison, W.T.A. *Microporous Mesoporous Mat.*, **23**, 197–202 (1998)
(22)   Chippindale, A.M., Cowley, A.R., Chen, J.S., Gao, Q. and Xu, R. *Acta Crystallogr.*, **C55**, 845–847 (1999)
(23)   Henry, P.F., Hughes, R.W., Ward, S.C. and Weller, M.T. *Chem. Commun.*, 1959–1960 (2000)
(24)   Henry, P.F., Weller, M.T. and Hughes, R.W. *Inorg. Chem.*, **39**, 5420–5421 (2000)
(25)   Krogh Andersen, I.G., Krogh Andersen, E., Norby, P., Colella, C. and Degennaro, M. *Zeolites*, **11**, 149–154 (1991)
(26)   Bu, X., Feng, P., Gier, T.E. and Stucky, G.D. *J. Solid State Chem.*, **136**, 210–215 (1998)

ACO

Framework Type Data

Im̄3m

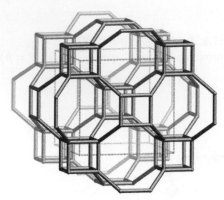 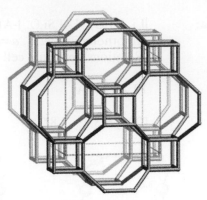

framework viewed along [001]

Idealized cell data: cubic, *Im̄3m*, a = 9.9Å

Coordination sequences and vertex symbols:

T₁ (16,3m)   4   9   19   35   52   72   100   131   163   201 $4·8_2·4·8_2·4·8_2$

Secondary building units: 8 or 4-4 or 4

Composite building units:
   d4r

Materials with this framework type:
   *ACP-1[1]

ACP-1   Type Material Data   **ACO**

Crystal chemical data: $|(C_2H_{10}N_2)_4 (H_2O)_2| [Al_{0.88}Co_{7.12}P_8O_{32}]$-**ACO**
           $C_2H_{10}N_2$ = ethylenediammonium
           tetragonal, $I\bar{4}2m$, $a = 10.240$Å, $c = 9.652$Å $^{(1)}$

Framework density: 15.8 T/1000Å³

Channels   <100> **8** 2.8 x 3.5** ↔ [001] **8** 3.5 x 3.5*

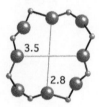 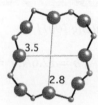 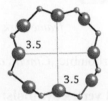 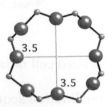

8-ring viewed along <100>   *8-ring viewed along [001]*

References:
(1) Feng, P., Bu, X. and Stucky, G.D. *Nature*, **388**, 735–741 (1997)

AEI

Framework Type Data

Cmcm

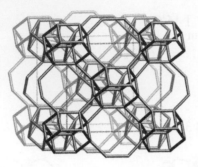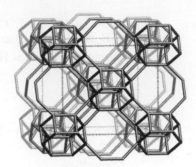

framework viewed along [001]

Idealized cell data: orthorhombic, *Cmcm*, a = 13.7Å, b = 12.6Å, c = 18.5Å

Coordination sequences and vertex symbols:

T_1 (16,1)	4	9	17	29	45	64	85	111	143	177	4·4·4·8·6·8
T_2 (16,1)	4	9	17	29	45	65	88	113	143	178	4·4·4·8·6·8
T_3 (16,1)	4	9	17	29	45	65	87	113	143	176	4·4·4·8·6·8

Secondary building units: 6-6 or 4-2 or 6 or 4

Composite building units:
   d6r

Materials with this framework type:
   *AlPO-18[1]
   [Co-Al-P-O]-**AEI**[2]
   SAPO-18[3]
   SIZ-8[4]
   SSZ-39[5]

AlPO-18   **Type Material Data**   **AEI**

Crystal chemical data: $[Al_{24}P_{24}O_{96}]$-**AEI**
monoclinic, $C2/c$

$a = 13.711$Å, $b = 12.732$Å, $c = 18.571$Å, $\beta = 90.01°^{(1)}$

Framework density: 14.8 T/1000Å^3

Channels: {[100] **8** 3.8 x 3.8 ↔ [110] **8** 3.8 x 3.8 ↔ [001] **8** 3.8 x 3.8}***

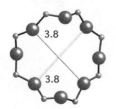 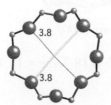 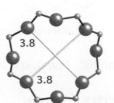 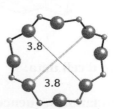

8-ring viewed along [100]   *8-ring viewed along [110]*

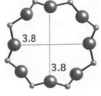 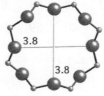

8-ring viewed along [001]

References:

(1) Simmen, A., McCusker, L.B., Baerlocher, Ch. and Meier, W.M. *Zeolites*, **11**, 654–661 (1991)
(2) Marchese, L., Chen, J.S., Thomas, J.M., Coluccia, S. and Zecchina, A. *J. Phys. Chem.*, **98**, 13350–13356 (1994)
(3) Chen, J.S., Thomas, J.M., Wright, P.A. and Townsend, R.P. *Catalysis Letters*, **28**, 241–248 (1994)
(4) Parnham, E.R. and Morris, R.E. *J. Am. Chem. Soc.*, **128**, 2204–2205 (2006)
(5) Wagner, P., Nakagawa, Y., Lee, G.S., Davis, M.E., Elomari, S., Medrud, R.C. and Zones, S.I. *J. Am. Chem. Soc.*, **122**, 263–273 (2000)

AEL

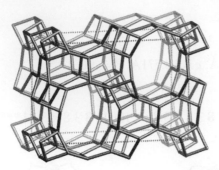
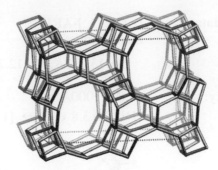

framework viewed along [100]

Idealized cell data: orthorhombic, *Imma*, *a* = 8.3Å, *b* = 18.7Å, *c* = 13.4Å

Coordination sequences and vertex symbols:

T_1 (16,1)	4	11	21	37	59	85	114	150	189	232	$4 \cdot 6_2 \cdot 6 \cdot 6_3 \cdot 6_2 \cdot 6_3$
T_2 (16,1)	4	11	22	38	58	85	115	148	188	234	$4 \cdot 6_2 \cdot 6 \cdot 6_3 \cdot 6_2 \cdot 6_3$
T_3 (8,*m*)	4	12	24	40	59	84	115	150	186	230	$6 \cdot 6_2 \cdot 6_2 \cdot 6_2 \cdot 6_2 \cdot 6_2$

Secondary building units: 4–1

Composite building units:

nsc   afi   bog

narsarsukite
chain

Materials with this framework type:
*AlPO-11[1,2]
GeAPO-11[3]
MnAPO-11[4]
SAPO-11 plus numerous compositional variants[5,6]

AlPO-11   **Type Material Data**   **AEL**

Crystal chemical data: $[Al_{20}P_{20}O_{80}]$-**AEL**
                                   orthorhombic, *Ibm*2, $a = 13.534$Å, $b = 18.482$Å, $c = 8.370$Å $^{(2)}$
                                   (Relationship to unit cell of Framework Type: $a' = c$, $b' = b$, $c' = a$)

Framework density: 19.1 T/1000Å^3

Channels: [001] **10** 4.0 x 6.5*

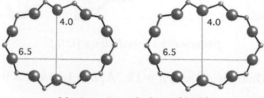

10-ring viewed along [001]

References:
(1) Bennett, J.M., Richardson Jr., J.W., Pluth, J.J. and Smith, J.V. *Zeolites*, **7**, 160–162 (1987)
(2) Richardson Jr., J.W., Pluth, J.J. and Smith, J.V. *Acta Crystallogr.*, **B44**, 367–373 (1988)
(3) Meriaudeau, P., Tuan, V.A., Hung, L.N. and Szabo, G. *Zeolites*, **19**, 449–451 (1997)
(4) Pluth, J.J., Smith, J.V. and Richardson Jr., J.W. *J. Phys. Chem.*, **92**, 2734–2738 (1988)
(5) Flanigen, E.M., Lok, B.M., Patton, R.L. and Wilson, S.T. *Pure Appl. Chem.*, **58**, 1351–1358 (1986)
(6) Flanigen, E.M., Lok, B.M., Patton, R.L. and Wilson, S.T. *Proc. 7th Int. Zeolite Conf.*, pp. 103–112 (1986)

AEN

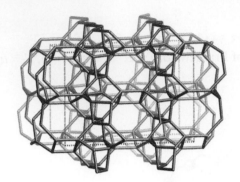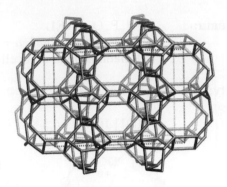

framework viewed along [001]

Idealized cell data: orthorhombic, *Cmce*, $a = 18.5$Å, $b = 13.4$Å, $c = 9.6$Å

Coordination sequences and vertex symbols:

T$_1$ (16,1)	4	11	24	41	60	86	123	162	199	248	$4 \cdot 6_3 \cdot 6_2 \cdot 8_2 \cdot 6_3 \cdot 8$
T$_2$ (16,1)	4	11	22	39	64	90	119	155	201	250	$4 \cdot 6 \cdot 6 \cdot 6_2 \cdot 6_2 \cdot 6_4$
T$_3$ (16,1)	4	11	22	38	63	90	116	155	204	250	$4 \cdot 6_3 \cdot 6_2 \cdot 6_3 \cdot 6_2 \cdot 8$

Secondary building units: 6 or 4

Materials with this framework type:

*AlPO-EN3[1]
[Ga-P-O]-**AEN**[2]
AlPO-53(A)[3]
AlPO-53(B)[3]
CFSAPO-1A[4]

CoIST-2[5]
IST-2[6]
JDF-2[7]
MCS-1[8]
MnAPO-14[9]

Mu-10[10]
UiO-12-500[11]
UiO-12-as[11]

AlPO-EN3   Type Material Data   **AEN**

Crystal chemical data: $|(C_2H_8N_2)_4 (H_2O)_{16}| [Al_{24}P_{24}O_{96}]$-**AEN**
$C_2H_8N_2$ = ethylenediamine
orthorhombic, $P2_12_12_1$, a = 10.292 Å, b = 13.636Å, c = 17.344 Å [1]
(Relationship to unit cell of Framework Type: a' = c, b' = b, c' = a)

Framework density: 19.7 T/1000Å^3

Channels: [100] **8** 3.1 x 4.3* ↔ [010] **8** 2.7 x 5.0*

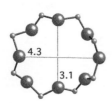 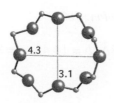  

         8-ring viewed along [100]   *8-ring viewed along [010]*

References:
(1) Parise, J.B. *Stud. Surf. Sci. Catal.*, **24**, 271–278 (1985)
(2) Glasser, F.P., Howie, R.A. and Kan, Q.B. *Acta Crystallogr.*, **C50**, 848–850 (1994)
(3) Kirchner, R.M., Grosse-Kunstleve, R.W., Pluth, J.J., Wilson, S.T., Broach, R.W. and Smith, J.V. *Microporous Mesoporous Mat.*, **39**, 319–332 (2000)
(4) He, H. and Long, Y. *J. Incl. Phenom.*, **5**, 591–599 (1987)
(5) Borges, C., Ribeiro, M.F., Henriques, C., Lourenco, J.P., Murphy, D.M., Louati, A. and Gabelica, Z. *J. Phys. Chem. B*, **108**, 8344–8354 (2004)
(6) Fernandes, A., Ribeiro, M.F., Borges, C., Lourenco, J.P., Rocha, J. and Gabelica, Z. *Microporous Mesoporous Mat.*, **90**, 112–128 (2006)
(7) Chippindale, A.M., Powell, A.V., Jones, R.H., Thomas, J.M., Cheetham, A.K., Huo, Q.S. and Xu, R.R. *Acta Crystallogr.*, **C50**, 1537–1540 (1994)
(8) Simmen, A. *Ph.D. Thesis, ETH, Zürich, Switzerland* (1992)
(9) Shi, L., Li, J.Y., Yu, J.H., Li, Y., Ding, H. and Xu, R. *Inorg. Chem.*, **43**, 2703–2707 (2004)
(10) Soulard, M., Patarin, J. and Marler, B. *Solid State Sci.*, **1**, 37–53 (1999)
(11) Kongshaug, K.O., Fjellvåg, H., Klewe, B. and Lillerud, K. P. *Microporous Mesoporous Mat.*, **39**, 333–339 (2000)

AET

Framework Type Data

Cmcm

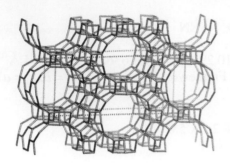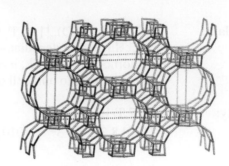

framework viewed along [001]

Idealized cell data: orthorhombic, *Cmcm*, $a = 32.8$Å, $b = 14.4$Å, $c = 8.4$Å

Coordination sequences and vertex symbols:

T_1 (16,1)	4	11	21	35	53	78	108	140	172	208	$4 \cdot 6_2 \cdot 6 \cdot 6_3 \cdot 6_2 \cdot 6_3$
T_2 (16,1)	4	11	21	35	52	74	102	136	172	212	$4 \cdot 6_2 \cdot 6_2 \cdot 6_3 \cdot 6_2 \cdot 6_3$
T_3 (16,1)	4	11	22	38	55	74	98	132	173	216	$4 \cdot 6_2 \cdot 6 \cdot 6_3 \cdot 6_2 \cdot 6_3$
T_4 (16,1)	4	12	23	36	52	75	103	135	172	215	$6 \cdot 6_2 \cdot 6_2 \cdot 6_2 \cdot 6_2 \cdot 6_2$
T_5 (8,*m*)	4	10	18	32	52	76	105	140	171	202	$4 \cdot 6_3 \cdot 4 \cdot 6_3 \cdot 6 \cdot 6_4$

Secondary building units: 6

Composite building units:

nsc	*dnc*	*afi*	*bog*
narsarsukite chain	*double narsarsukite chain*		

Materials with this framework type:
   *AlPO-8[1,2]
   MCM-37[3]

AlPO-8   **Type Material Data**   **AET**

Crystal chemical data: [Al$_{36}$P$_{36}$O$_{144}$]-**AET**
                              orthorhombic, *Cmc*2$_1$, *a* = 33.29Å, *b* = 14.76Å, *c* = 8.257Å [1]

Framework density: 17.7 T/1000Å^3

Channels: [001] **14** 7.9 x 8.7*

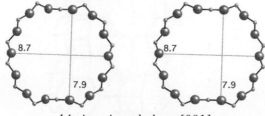

14-ring viewed along [001]

References:
(1) Dessau, R.M., Schlenker, J.L. and Higgins, J.B. *Zeolites*, **10**, 522–524 (1990)
(2) Richardson Jr., J.W. and Vogt, E.T.C. *Zeolites*, **12**, 13-19 (1992)
(3) Chu, C.T.W., Schlenker, J.L., Lutner, J.D. and Chang, C.D. *U.S. Patent 5,091,073* (1992)

AFG

Framework Type Data

$P6_3/mmc$

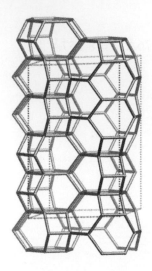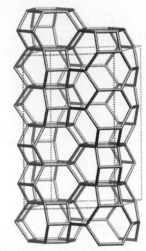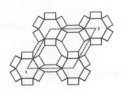

framework viewed normal to [001]

Idealized cell data: hexagonal, $P6_3/mmc$, $a = 12.5$Å, $c = 20.8$Å

Coordination sequences and vertex symbols:

T_1 (24,1)	4	10	20	34	53	76	103	135	170	208	4·6·4·6·6·6
T_2 (12,m)	4	10	20	34	54	78	104	134	168	210	4·6·4·6·6·6
T_3 (12,2)	4	10	20	34	54	78	104	134	168	210	4·4·6·6·6·6

Secondary building units: 6 or 4

Framework description: ABABACAC sequence of 6-rings

Composite building units:

can   *lio*

Materials with this framework type:
*Afghanite[1-4]

| Afghanite | Type Material Data | **AFG** |

Crystal chemical data: |Ca$_{9.8}$Na$_{22}$ (H$_2$O)$_4$ Cl$_2$(SO$_4$)$_{5.3}$CO$_3$| [Al$_{24}$Si$_{24}$O$_{96}$]-**AFG**
hexagonal, $P6_3mc$, $a = 12.761$Å, $c = 21.416$Å $^{(3)}$

Framework density: 15.9 T/1000Å^3

Channels: apertures formed by 6-rings only

References:
(1) Bariand, P., Cesbron, F. and Giraud, R. *Bull. Soc. fr. Minéral. Cristallogr.*, **91**, 34–42 (1968)
(2) Merlino, S. and Mellini, M. *Zeolite 1976, Program and Abstracts, Tucson* (1976)
(3) Pobedimskaya, E.A., Rastsvetaeva, R.K., Terent'eva, L.E. and Sapozhnikov, A.N. *Dokl. Akad. Nauk SSSR*, **320**, 882–886 (1991)
(4) Ballirano, P., Bonaccorsi, E., Maras, A. and Merlino, S., *Eur. J. Mineral.*, **9**, 21–30 (1997)

AFI

Framework Type Data

P6/mcc

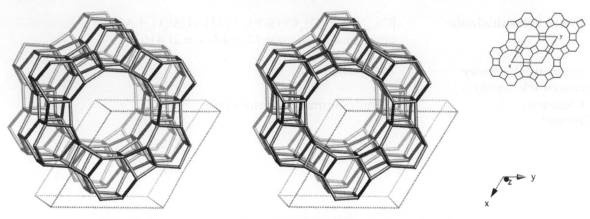

framework viewed along [001] (upper right: projection down [001])

Idealized cell data: hexagonal, *P6/mcc*, $a = 13.8$Å, $c = 8.6$Å

Coordination sequences and vertex symbols:

| T_1 (24,1) | 4 | 11 | 21 | 35 | 53 | 77 | 105 | 137 | 172 | 212 |

$4 \cdot 6_2 \cdot 6 \cdot 6_3 \cdot 6_2 \cdot 6_3$

Secondary building units: 12 or 6 or 4

Composite building units:

nsc	afi	bog

narsarsukite
chain

Materials with this framework type:

*AlPO-5[1]
[Sn-Al-P-O]-**AFI**[2]
|(Ni(deta)$_2$)-|[Al-P-O-F-]-**AFI**[3]
|TPAF-|[Al-P-O]-**AFI**[4]
CoAPO-5[5,6]
CrAPO-5[7]
FAPO-5[8]

MAPO-5, M = Cd, Cu, Mo, V/Mo, Zr[9]
MAPO-5, M=Mg, Mn[10]
SAPO-5 and numerous compositional variants[11,12]
SSZ-24[13]
VAPO-5[14]
ZnAPO-5[15]

AlPO-5 **Type Material Data** **AFI**

Crystal chemical data: $|(C_{12}H_{28}N)(H_2O)_x(OH)|[Al_{12}P_{12}O_{48}]$-**AFI**
$C_{12}H_{28}N$ = tetrapropylammonium
hexagonal, $P6cc$, $a = 13.726$Å, $c = 8.484$Å [1]

Framework density: 17.3 T/1000Å^3

Channels: [001] **12** 7.3 x 7.3*

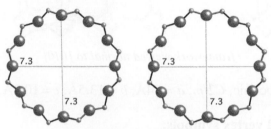

12-ring viewed along [001]

References:
(1) Bennett, J.M., Cohen, J.P., Flanigen, E.M., Pluth, J.J. and Smith, J.V. *ACS Sym. Ser.*, **218**, 109–118 (1983)
(2) Flavell, W.R., Nicholson, D.G., Nilsen, M.H. and Ståhl, K. *J. Mater. Chem.*, **11**, 620–627 (2001)
(3) Garcia, R., Shannon, I.J., Slawin, A.M.Z., Zhou, W., Cox, P.A. and Wright, P.A. *Microporous Mesoporous Mat.*, **58**, 91–104 (2003)
(4) Qiu, S., Pang, W., Kessler, H. and Guth, J.L. *Zeolites*, **9**, 440–444 (1989)
(5) Montes, C., Davis, M.E., Murray, B. and Narayana, M. *J. Phys. Chem.*, **94**, 6425–6430 (1990)
(6) Chao, K.J., Sheu, S.P. and Sheu, H.S. *J. Chem. Soc., Faraday Trans.*, **88**, 2949–2954 (1992)
(7) Radaev, S., Joswig, W. and Baur, W.H. *J. Mater. Chem.*, **6**, 1413–1418 (1996)
(8) Zenonos, C., Sankar, G., Cora, F., Lewis, D.W., Pankhurst, Q.A., Catlow, C.R.A. and Thomas, J.M. *Phys Chem Chem Phys*, **4**, 5421–5429 (2002)
(9) Kornatowski, J., Sychev, M., Finger, G., Baur W.H., Rozwadowski, M. and Zibrowius, B. *Proc. Polish-German Zeolite Colloquium*, 20–26 (1992)
(10) Parrillo, D.J., Pereira, C., Kokotailo, G.T. and Gorte, R.J. *J. Catal.*, **138**, 377–385 (1992)
(11) Flanigen, E.M., Lok, B.M., Patton, R.L. and Wilson, S.T. *Pure Appl. Chem.*, **58**, 1351–1358 (1986)
(12) Flanigen, E.M., Lok, B.M., Patton, R.L. and Wilson, S.T. *Proc. 7th Int. Zeolite Conf.*, pp. 103–112 (1986)
(13) Bialek, R., Meier, W.M., Davis, M. and Annen, M.J. *Zeolites*, **11**, 438–442 (1991)
(14) Bedioui, F., Briot, E., Devynck, J. and Balkus, K.J. *Inorganica Chimica Acta*, **254**, 151–155 (1997)
(15) Christensen, A.N. and Hazell, R.G. *Acta Chemica Scand.*, **53**, 403–409 (1999)

AFN

Framework Type Data

C2/m

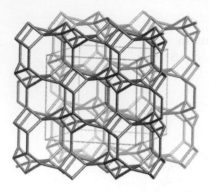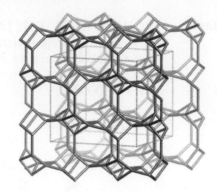

framework viewed normal to [010]

Idealized cell data: monoclinic, $C2/m$, $a = 14$Å, $b = 13.5$Å, $c = 10.2$Å, $\beta = 107.2°$

Coordination sequences and vertex symbols:

T_1 (8,1)	4	9	19	33	51	76	98	123	162	203		$4·4·4·8_2·6·8_4$
T_2 (8,1)	4	9	18	31	49	72	99	130	160	198		$4·4·4·8_2·6·6_2$
T_3 (8,1)	4	9	17	30	49	75	102	125	157	202		$4·6·4·8·4·8_7$
T_4 (8,1)	4	10	21	35	50	71	100	132	164	198		$4·4·6·8·8·8_2$

Secondary building units: 8 or 4

Composite building units:
mei

Materials with this framework type:
　　*AlPO-14[1,2]
　　|(C_3N_2H_12)-|[Mn-Al-P-O]-**AFN**[3]
　　GaPO-14[4]

AlPO-14   **Type Material Data**   **AFN**

Crystal chemical data: $[Al_8P_8O_{32}]$-**AFN**
                              triclinic, $P\bar{1}$, $a = 9.704$Å, $b = 9.736$Å, $c = 10.202$Å
                              $\alpha = 77.81°$, $\beta = 77.50°$, $\gamma = 87.69°$ [1]

Framework density: 17.4 T/1000Å^3
Channels: [100] **8** 1.9 x 4.6* ↔ [010] **8** 2.1 x 4.9* ↔ [001] **8** 3.3 x 4.0*

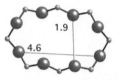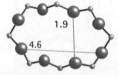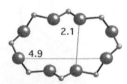

8-ring viewed along [100]   *8-ring viewed along [010]*

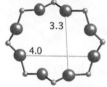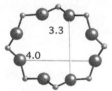

8-ring viewed along [001]

References:
(1) Broach, R.W., Wilson, S.T. and Kirchner, R.M. *Proc. 12th Int. Zeolite Conf.*, **III**, pp. 1715–1722 (1999)
(2) Broach, R.W., Wilson, S.T. and Kirchner, R.M. *Microporous Mesoporous Mat.*, **57**, 211–214 (2003)
(3) Shi, L., Li, J., Duan, F., Yu, J., Li, Y. and Xu, R. *Microporous Mesoporous Mat.*, **85**, 252–259 (2005)
(4) Parise, J.B. *Acta Crystallogr.*, **C42**, 670–673 (1986)

AFO

Framework Type Data

Cmcm

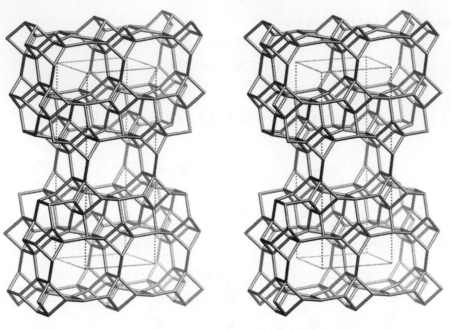

framework viewed along [001]

Idealized cell data: orthorhombic, *Cmcm*, a = 9.8Å, b = 25.6Å, c = 8.3Å

Coordination sequences and vertex symbols:

T$_1$ (16,1)	4	11	22	38	58	85	115	149	190	235	4·6$_2$·6·6$_3$·6$_2$·6$_3$
T$_2$ (8,m)	4	11	22	41	65	88	111	145	186	231	4·6$_2$·6·6$_3$·6·6$_3$
T$_3$ (8,m)	4	11	21	36	56	82	115	156	195	231	4·6$_2$·6$_2$·6$_3$·6$_2$·6$_3$
T$_4$ (8,m)	4	12	23	37	55	82	118	155	189	232	6·6$_2$·6$_2$·6$_2$·6$_2$·6$_2$

Secondary building units: 2-6-2 or 4-1

Composite building units:

nsc	*afi*	*bog*
narsarsukite chain		

Materials with this framework type:

*AlPO-41[1]   MnAPSO-41[2]

MnAPO-41[2]   SAPO-41[2]

AlPO-41 **Type Material Data** **AFO**

Crystal chemical data: $[Al_{10}P_{10}O_{40}]$-**AFO**
monoclinic, $P2_1$

$a =$ 9.718Å, $b =$ 13.792Å, $c =$ 8.359Å, $\gamma =$ 110.6°[1]
(Relationship to unit cell of Framework Type:

$a' = a$, $b' = b/(2\sin\gamma')$, $c' = c$
or, as vectors, $\mathbf{a'} = \mathbf{a}$, $\mathbf{b'} = (\mathbf{b} - \mathbf{a})/2$, $\mathbf{c'} = \mathbf{c}$)

Framework density: 19.1 T/1000Å^3

Channels: [001] **10** 4.3 x 7.0*

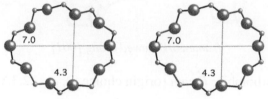

10-ring viewed along [001]

References:
(1) Kirchner, R.M. and Bennett, J.M. *Zeolites*, **14**, 523–528 (1994)
(2) Hartmann, M., Prakash, A.M. and Kevan, L. *J. Chem. Soc., Faraday Trans.*, **94**, 723–727 (1998)

AFR

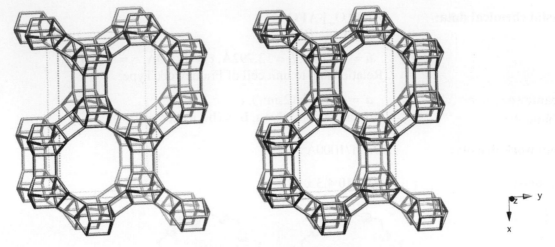

framework viewed along [001]

Idealized cell data: orthorhombic, *Pmmn* (origin choice 2), a = 22.3Å, b = 13.6Å, c = 7Å

Coordination sequences and vertex symbols:

T_1 (8,1)	4	9	16	27	43	63	88	115	141	171		$4·6·4·6_2·4·8$
T_2 (8,1)	4	9	18	30	43	64	90	111	140	181		$4·4·4·8·6_3·8$
T_3 (8,1)	4	9	18	29	42	66	93	112	139	177		$4·4·4·12·6·6_3$
T_4 (8,1)	4	10	17	28	47	65	86	117	144	169		$4·6·4·6·6·12$

Secondary building units: 6-2 or 4-4- or 4

Composite building units:
sti

Materials with this framework type:

*SAPO-40[1-3]
AlPO-40[4,5]
CoAPO-40[4]

CoAPSO-40[6]
ZnAPO-40[4]
ZnAPSO-40[6]

SAPO-40   **Type Material Data**   **AFR**

Crystal chemical data: $|(C_{12}H_{28}N)_4\ (OH)_4|\ [Si_8Al_{28}P_{28}\ O_{128}]$-**AFR**

                                $C_{12}H_{28}N$ = tetrapropylammonium

                             orthorhombic, *Pccn*, a = 21.944Å, b = 13.691Å, c = 14.249Å [3]

                             (Relationship to unit cell of Framework Type: a' = a, b' = b, c' = $2c$)

Framework density: 15 T/1000Å^3

Channels: [001] **12** 6.7 x 6.9* ↔ [010] **8** 3.7 x 3.7*

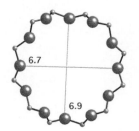 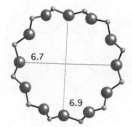 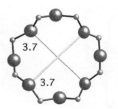 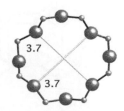

12-ring viewed along [001]   *8-ring viewed along [010]*

References:
(1) Estermann, M.A., McCusker, L.B. and Baerlocher, Ch. *J. Appl. Crystallogr.*, **25**, 539–543 (1992)
(2) Dumont, N., Gabelica, Z., Derouane, E.G. and McCusker, L.B. *Microporous Materials*, **1**, 149–160 (1993)
(3) McCusker, L.B. and Baerlocher, Ch. *Microporous Materials*, **6**, 51–54 (1996)
(4) Lourenco, J.P., Ribeiro, M.F., Ribeiro, F.R., Rocha, J., Onida, B., Garrone, E. and Gabelica, Z. *Zeolites*, **18**, 398–407 (1997)
(5) Ramaswamy, V., McCusker, L.B. and Baerlocher, Ch. *Microporous Mesoporous Mat.*, **31**, 1–8 (1999)
(6) Lourenco, J.P., Ribeiro, M.F., Borges, C., Rocha, J., Onida, B., Garrone, E. and Gabelica, Z. *Microporous Mesoporous Mat.*, **38**, 267–278 (2000)

AFS

Framework Type Data

P6₃/mcm

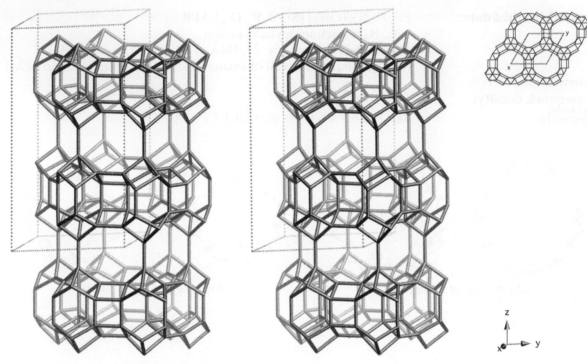

framework viewed normal to [001] (upper right: projection down [001])

Idealized cell data: hexagonal, $P6_3/mcm$, $a = 13.1$Å, $c = 25.9$Å

Coordination sequences and vertex symbols:

T₁ (24,1)	4	9	17	28	42	60	83	111	138	166		4·4·4·8₂·6₂·8
T₂ (24,1)	4	9	16	25	39	61	86	109	134	163		4·4·4·6· 6·12
T₃ (8,3)	4	9	18	30	43	62	85	105	135	180		4·8·4·8·4·8

Secondary building units: 6*1

Composite building units:

 afs   bph

Materials with this framework type:
   *MAPSO-46[1]
   MAPO-46[2]

MAPSO-46   **Type Material Data**   **AFS**

Crystal chemical data: |(C$_6$H$_{16}$N)$_8$ (H$_2$O)$_{14}$| [Mg$_6$Al$_{22}$P$_{26}$Si$_2$O$_{112}$]-**AFS**
C$_6$H$_{16}$N = dipropylammonium
trigonal, $P3c1$, a = 13.225Å, c = 26.892Å [1]

Framework density: 13.7 T/1000Å^3

Channels: [001] **12** 7.0 x 7.0* ↔ ⊥ [001] **8** 4.0 x 4.0**

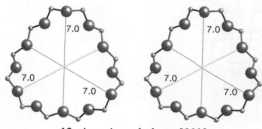 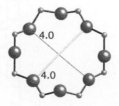 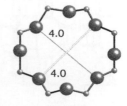

12-ring viewed along [001]   *8-ring viewed normal to [001]*

References:
(1) Bennett, J.M. and Marcus, B.K. *Stud. Surf. Sci. Catal.*, **37**, 269–279 (1988)
(2) Akolekar, D.B. and Kaliaguine, S. *J. Chem. Soc., Faraday Trans.*, **89**, 4141–4147 (1993)

AFT

Framework Type Data

$P6_3/mmc$

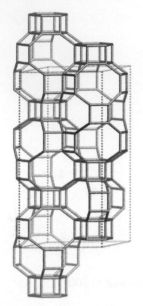 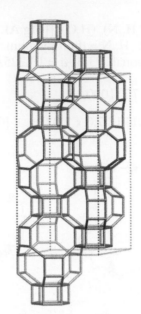 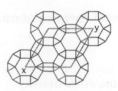

framework viewed normal to [001] (upper right: projection down [001])

Idealized cell data: hexagonal, $P6_3/mmc$, $a = 13.7$Å, $c = 29.4$Å

Coordination sequences and vertex symbols:

T_1 (24,1)	4	9	17	29	45	64	85	110	140	173		4·4·4·8·6·8
T_2 (24,1)	4	9	17	29	45	64	86	113	144	178		4·4·4·8·6·8
T_3 (24,1)	4	9	17	29	45	65	88	113	141	175		4·4·4·8·6·8

Secondary building units: 6-6 or 4-2 or 6 or 4

Framework description: AABBCCAACCBB sequence of 6-rings

Composite building units:

d6r	*gme*	*cha*	*aft*

Materials with this framework type:
*AlPO-52[1,2]

AlPO-52   **Type Material Data**   # AFT

Crystal chemical data: $[Al_{36}P_{36}O_{144}]$-**AFT**
trigonal, $P\bar{3}1c$, $a = 13.715$Å, $c = 29.676$Å $^{(2)}$

Framework density: 14.9 T/1000Å^3

Channels: $\perp$ [001] **8** 3.2 x 3.8***

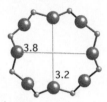 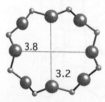 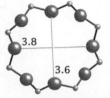 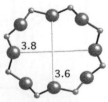

gme cage 8-ring viewed normal to [001]   *cha cage 8-ring viewed normal to [001]*

References:
(1) Bennett, J.M., Kirchner, R.M. and Wilson, S.T. *Stud. Surf. Sci. Catal.*, **49**, 731–739 (1989)
(2) McGuire, N.K., Bateman, C.A., Blackwell, C.S., Wilson, S.T. and Kirchner, R.M. *Zeolites*, **15**, 460–469 (1995)

AFX

Framework Type Data

P6₃/mmc

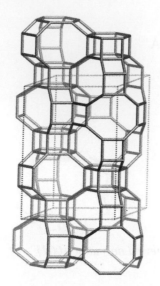 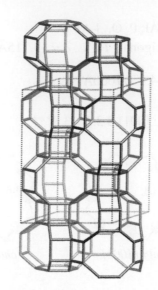 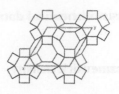

framework viewed normal to [001] (upper right: projection down [001])

Idealized cell data: hexagonal, $P6_3/mmc$, $a = 13.7$Å, $c = 19.7$Å

Coordination sequences and vertex symbols:

T₁ (24,1)	4	9	17	29	45	65	89	116	144	175	4·4·4·8·6·8
T₂ (24,1)	4	9	17	29	45	64	85	110	141	178	4·4·4·8·6·8

Secondary building units: 6-6 or 4-2 or 6 or 4

Framework description: AABBCCBB sequence of 6-rings

Composite building units:

d6r   gme   aft

Materials with this framework type:
  *SAPO-56[1,2]
  MAPSO-56, M=Co, Mn, Zr[3]
  SSZ-16[4]

SAPO-56   Type Material Data   **AFX**

Crystal chemical data: $|H_3|$ $[Si_5Al_{23}P_{20}O_{96}]$-**AFX**
trigonal, $P\bar{3}1c$, $a = 13.762$Å, $c = 19.949$Å [2]

Framework density: 14.7 T/1000Å^3

Channels: $\perp$ [001] **8** 3.4 x 3.6***

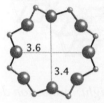

8-ring viewed normal to [001]

References:
(1) McGuire, N.K., Blackwell, C.S., Bateman, C.A., Wilson, S.T. and Kirchner, R.M. *private communication*
(2) Wilson, S.T., Broach, R.W., Blackwell, C.S., Bateman, C.A., McGuire, N.K. and Kirchner, R.M. *Microporous Mesoporous Mat.*, **28**, 125–137 (1999)
(3) Tian, P., Liu, Z., Xu, L. and Sun, C. *Stud. Surf. Sci. Catal.*, **135**, 248 (05-P-18) (2001)
(4) Lobo, R.F., Zones, S.I. and Medrud, R.C. *Chem. Mater.*, **8**, 2409–2411 (1996)

AFY

Framework Type Data

$P\bar{3}1m$

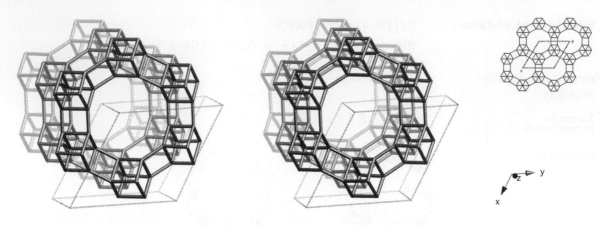

framework viewed along [001] (upper right: projection down [001])

Idealized cell data: trigonal, $P\bar{3}1m$, $a = 12.3$Å, $c = 8.6$Å

Coordination sequences and vertex symbols:

T$_1$ (12,1)	4	8	14	25	39	53	71	96	124	152	4·4·4·8·4·12
T$_2$ (4,3)	4	9	16	23	34	57	82	98	115	141	4·8·4·8·4·8

Secondary building units: 4-4 or 4

Composite building units:
    d4r

Materials with this framework type:
    *CoAPO-50[1,2]
    MgAPO-50[3]
    MnAPO-50[4]
    ZnAPO-50[5]

CoAPO-50 **Type Material Data** **AFY**

Crystal chemical data: $|(C_6H_{16}N)_3 (H_2O)_7| [Co_3Al_5P_8O_{32}]$-**AFY**
$C_6H_{16}N$ = dipropylammonium
trigonal, $P\bar{3}$, a = 12.747Å, c = 9.015Å [2]

Stability: Unstable to removal of template [1]

Framework density: 12.6 T/1000Å^3

Channels: [001] **12** 6.1 x 6.1* ↔ ⊥ [001] **8** 4.0 x 4.3**

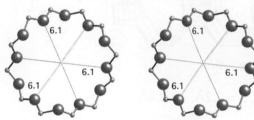
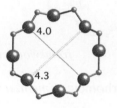
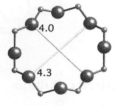

12-ring viewed along [001]   8-ring viewed normal to [001]

References:
(1) Wilson, S.T. private communication
(2) Bennett, J.M. and Marcus, B.K. Stud. Surf. Sci. Catal., **37**, 269–279 (1988)
(3) Akolekar, D.B. Zeolites, **15**, 583–590 (1995)
(4) Tusar, N.N., Ristic, A., Meden, A. and Kaucic, V. Microporous Mesoporous Mat., **37**, 303–311 (2000)
(5) Arcon, I., Tusar, N.N., Ristic, A., Kaucic, V., Kodre, A. and Helliwell, M. J. Synch. Rad., **8**, 590–592 (2001)

AHT

Framework Type Data

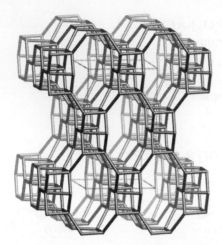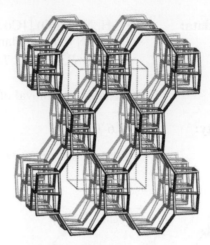

framework viewed along [001]

Idealized cell data: orthorhombic, *Cmcm*, $a = 15.8$Å, $b = 9.2$Å, $c = 8.6$Å

Coordination sequences and vertex symbols:

T_1 (16,1)	4	11	21	36	56	81	109	142	179	221	$4 \cdot 6_2 \cdot 6 \cdot 6_3 \cdot 6 \cdot 6_3$
T_2 (8,*m*)	4	10	18	32	53	78	105	140	179	218	$4 \cdot 6_3 \cdot 4 \cdot 6_3 \cdot 6 \cdot 6_4$

Secondary building units: 4-2 or 6

Composite building units:
        dnc   *bog*

        double
     narsarsukite
        chain

Materials with this framework type:
*AlPO-H2[1-3]

AlPO-H2 **Type Material Data** **AHT**

Crystal chemical data:	$\|(H_2O)_8\|$ $[Al_6P_6O_{24}]$-**AHT**
	monoclinic, $P2_1$, $a = 9.486$Å, $b = 9.914$Å, $c = 8.126$Å, $\gamma = 121.49°$[(1)]
	(Relationship to unit cell of Framework Type:
	$a' = a/(2\sin\gamma')$, $b' = b$, $c' = c$
	or, as vectors, $\mathbf{a}' = (\mathbf{a} - \mathbf{b})/2$, $\mathbf{b}' = \mathbf{b}$, $\mathbf{c}' = \mathbf{c}$)
Stability:	Transforms to AlPO$_4$-tridymite on heating[(3)]
Framework density:	18.4 T/1000Å^3
Channels:	[001] **10** 3.3 x 6.8*

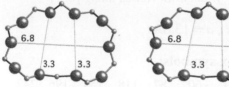

10-ring viewed along [001]

References:
(1) Higgins, J.B. private communication
(2) Li, H.X., Davis, M.E., Higgins, J.B. and Dessau, R.M. Chem. Commun., 403–405 (1993)
(3) Kennedy, G.J., Higgins, J.B., Ridenour, C.F., Li, H.X. and Davis, M.E. Solid State Nucl. Mag. Res., **4**, 173–178 (1995)

ANA   Framework Type Data   *Ia̅3d*

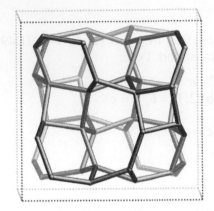 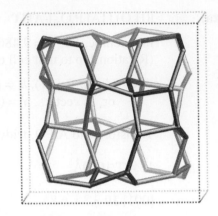

framework viewed along [001]

Idealized cell data: cubic, *Ia̅3d* , $a = 13.6$Å

Coordination sequences and vertex symbols:

T_1 (48,2)	4	10	22	39	60	87	118	154	196	242

$4 \cdot 4 \cdot 6 \cdot 6 \cdot 8_4 \cdot 8_4$

Secondary building units: 6-2 or 6 or 4-[1,1] or 1-4-1 or 4

Materials with this framework type:

*Analcime[1-3]
[Al-Co-P-O]-**ANA**[4]
[Al-Si-P-O]-**ANA**[5]
[Ga-Ge-O]-**ANA**[6]
[Zn-As-O]-**ANA**[7]
|(NH₄)-|[Be-B-P-O]-**ANA**[8]
|(NH₄)-|[Zn-Ga-P-O]-**ANA**[9]
|Cs-|[Al-Ge-O]-**ANA**[10]
|Cs-|[Be-Si-O]-**ANA**[11]
|Cs-Fe|[Si-O]-**ANA**[12]

|Cs-Na-(H₂O)|[Ga-Si-O]-**ANA**[13]
|Cs₁₆|[Cu₈Si₄₀O₉₆]-**ANA**[14]
|K-|[Be-B-P-O]-**ANA**[15]
|K-|[B-Si-O]-**ANA**[16]
|Li-|[Li-Zn-Si-O]-**ANA**[17]
|Li-Na|[Al-Si-O]-**ANA**[18]
|Na-|[Be-B-P-O]-**ANA**[19]
AlPO-24[20]
AlPO₄-pollucite[21]
Ammonioleucite[22]
Ca-D[23]

Hsianghualite[24]
Leucite[25]
Na-B[26]
Pollucite[27]
Synthetic analcime[28]
Synthetic hsinghualite[29]
Synthetic wairakite[30]
Wairakite, compositional variants[31]

References:
(1) Taylor, W.H. *Z. Kristallogr.*, **74**, 1–19 (1930)
(2) Knowles, C.R., Rinaldi, F.F. and Smith, J.V. *Indian Mineral.*, **6**, 127- (1965)
(3) Ferraris, G., Jones, D.W. and Yerkess, J. *Z. Kristallogr.*, **135**, 240–252 (1972)
(4) Feng, P., Bu, X. and Stucky, G.D. *Nature*, **388**, 735–741 (1997)
(5) Artioli, G., Pluth, J.J. and Smith, J.V. *Acta Crystallogr.*, **C40**, 214–217 (1984)
(6) Bu, X., Feng, P., Gier, T.E., Zhao, D. and Stucky, G.D. *J. Am. Chem. Soc.*, **120**, 13389–13397 (1998)
(7) Feng, P., Zhang, T. and Bu, X. *J. Am. Chem. Soc.*, **123**, 8608–8609 (2001)
(8) Zhang, H.Y., Chen, Z.X., Weng, L.H., Zhou, Y.M. and Zhao, D.Y. *Microporous Mesoporous Mat.*, **57**, 309–316 (2003)

Analcime   Type Material Data   ANA

Crystal chemical data: $|Na_{16} (H_2O)_{16}| [Al_{16}Si_{32}O_{96}]$-**ANA**
cubic, $Ia\overline{3}d$, $a = 13.73\text{Å}$ [3]

Framework density: $18.5 \text{ T}/1000\text{Å}^3$

Channels: irregular channels formed by highly distorted 8-rings

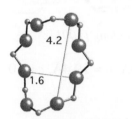 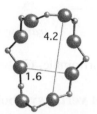

distorted 8-ring viewed along [110]

References (cont.):

(9)   Logar, N.Z., Mrak, M., Kaucic, V. and Golobic, A. *J. Solid State Chem.*, **156**, 480–486 (2001)

(10) Tripathi, A. and Parise, J.B. *Microporous Mesoporous Mat.*, **52**, 65–78 (2002)

(11) Torres-Martines, L.M., Gard, J.A., Howie, R.A. and West, A.R. *J. Solid State Chem.*, **51**, 100–103 (1984)

(12) Kopp, O.C., Harris, L.A., Clark, G.W. and Yakel, H.L. *Am. Mineral.*, **48**, 100–109 (1963)

(13) Yelon, W.B., Xie, D., Newsam, J.M. and Dunn, J. *Zeolites*, **10**, 553–558 (1990)

(14) Heinrich, A.R. and Baerlocher, Ch. *Acta Crystallogr.*, **C47**, 237–241 (1991)

(15) Zhang, H.Y., Chen, Z.X., Weng, L.H., Zhou, Y.M. and Zhao, D.Y. *Microporous Mesoporous Mat.*, **57**, 309–316 (2003)

(16) Millini, R., Montanari, L. and Bellussi, G. *Microporous Materials*, **1**, 9–15 (1993)

(17) Park, S.H., Gies, H., Toby, B.H. and Parise, J.B. *Chem. Mater.*, **14**, 3187–3196 (2002)

(18) Seretkin, Y.V., Bakakin, V.V. and Bazhan, I.S. *J. Struct. Chem.*, **46**, 659–671 (2005)

(19) Zhang, H.Y., Chen, Z.X., Weng, L.H., Zhou, Y.M. and Zhao, D.Y. *Microporous Mesoporous Mat.*, **57**, 309–316 (2003)

(20) Wilson, S.T., Lok, B.M., Messina, C.A., Cannan, T.R. and Flanigen, E.M. *J. Am. Chem. Soc.*, **104**, 1146–1147 (1982)

(21) Keller, E.B. *Ph.D. Thesis, ETH, Zürich, Switzerland* (1987)

(22) Hori, H., Nagashima, K., Yamada, M., Miyawaki, R. and Marubashi, T. *Am. Mineral.*, **71**, 1022–1027 (1986)

(23) Ames, L.L. and Sand, L.B. *Am. Mineral.*, **43**, 476–480 (1958)

(24) Wen-Hui, H., Saho-Hua, T., Kung-Hai, W., Chun-Lin, C. and Cheng Chi, Y. *Am. Mineral.*, **44**, 1327–1328 (1959)

(25) Peacor, D.R. *Z. Kristallogr.*, **127**, 213–224 (1968)

(26) Barrer, R.M. and White, E.A.D. *J. Chem. Soc.*, 1561–1571 (1952)

(27) Nel, H.J. *Am. Mineral.*, **29**, 443–451 (1944)

(28) Ghobarkar, H. and Franke, W. *Cryst. Res. Technol.*, 1071–1075 (1986)

(29) Ghobarkar, H., Schaef, O. and Knauth, Pl *Annal. Chimie, Science Matériaux*, **24**, 209–215 (1999)

(30) Ghobarkar, H. *Cryst. Res. Technol.*, K90–92 (1985)

(31) Takeuchi, Y., Mazzi, F., Haga, N. and Galli, E. *Am. Mineral.*, **64**, 993–1001 (1979)

APC

Framework Type Data

Cmce

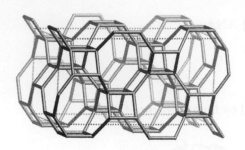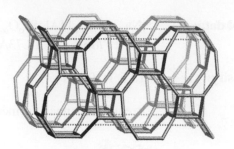

z
x y

framework viewed along [100]

Idealized cell data: orthorhombic, *Cmce*, $a = 9.0$Å, $b = 19.4$Å, $c = 10.4$Å

Coordination sequences and vertex symbols:

T_1 (16,1)	4	9	19	35	53	75	102	132	168	208	$4 \cdot 4 \cdot 4 \cdot 8_2 \cdot 8 \cdot 8_2$
T_2 (16,1)	4	10	20	35	54	76	104	136	171	211	$4 \cdot 6 \cdot 4 \cdot 6 \cdot 6 \cdot 8_2$

Secondary building units: 8 or 4

Composite building units:
   dcc
   double
   crankshaft chain

Materials with this framework type:
   *AlPO-C[1,2]
   AlPO-H3[3]
   CoAPO-H3[4]

AlPO-C   **Type Material Data**   **APC**

Crystal chemical data: $[Al_{16}P_{16}O_{64}]$-**APC**
orthorhombic, *Pbca*, $a = 19.821$Å, $b = 10.028$Å, $c = 8.936$Å [2]
(Relationship to unit cell of Framework Type: $a' = b$, $b' = c$, $c' = a$)

Stability: Transforms to AlPO-D at ca 250°C [2]

Framework density: 18 T/1000Å^3

Channels: [001] **8** 3.4 x 3.7* ↔ [100] **8** 2.0 x 4.7*

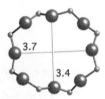 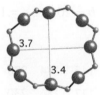 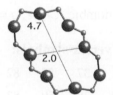 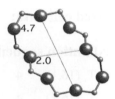

         8-ring viewed along [001]   *distorted 8-ring viewed along [100]*

References:
(1) Bennett, J.M., Dytrych, W.J., Pluth, J.J., Richardson Jr., J.W. and Smith, J.V. *Zeolites*, **6**, 349–359 (1986)
(2) Keller, E.B., Meier, W.M. and Kirchner, R.M. *Solid State Ionics*, **43**, 93–102 (1990)
(3) Pluth, J.J. and Smith, J.V. *Acta Crystallogr.*, **C42**, 1118–1120 (1986)
(4) Canesson, L., Arcon, I., Caldarelli, S. and Tuel, A. *Microporous Mesoporous Mat.*, **26**, 117–131 (1998)

APD

Framework Type Data

Cmce

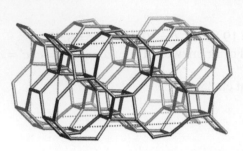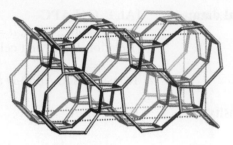

framework viewed along [100]

Idealized cell data: orthorhombic, *Cmce*, $a = 8.7$Å, $b = 20.1$Å, $c = 10.2$Å

Coordination sequences and vertex symbols:

T$_1$ (16,1)	4	10	21	37	57	82	112	145	184	228	$4 \cdot 4 \cdot 6_2 \cdot 8_3 \cdot 6_3 \cdot 8_3$
T$_2$ (16,1)	4	11	22	38	59	83	113	147	186	230	$4 \cdot 6_2 \cdot 6 \cdot 6_2 \cdot 6 \cdot 6_3$

Secondary building units: 8 or 6-2 or 4

Composite building units:
>   *nsc*
>   *narsarsukite*
>   *chain*

Materials with this framework type:
   *AlPO-D[1]
   APO-CJ3[2]

AlPO-D   Type Material Data   **APD**

Crystal chemical data: $[Al_{16}P_{16}O_{64}]$-**APD**

> (forms irreversibly from AlPO-C at around 200°C)
> orthorhombic, $Pca2_1$, $a = 19.187$Å, $b = 8.576$Å, $c = 9.804$Å [1]
> (Relationship to unit cell of Framework Type: $a' = b$, $b' = a$, $c' = c$)

Framework density: 19.8 T/1000Å^3

Channels: [010] **8** 2.3 x 6.0* ↔ [201] **8** 1.3 x 5.8*

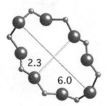 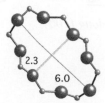   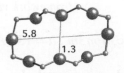 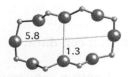

distorted 8-ring viewed along [010]   *distorted 8-ring along [201]*

References:
(1) Keller, E.B., Meier, W.M. and Kirchner, R.M. *Solid State Ionics*, **43**, 93–102 (1990)
(2) Wang, K.X., Yu, J.H., Zhu, G.S., Zou, Y.C. and Xu, R.R. *Microporous Mesoporous Mat.*, **39**, 281–289 (2000)

AST

Framework Type Data

Fm$\bar{3}$m

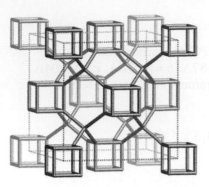 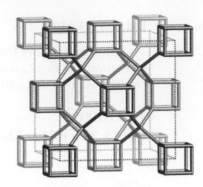

framework viewed along [100]

Idealized cell data: cubic, *Fm$\bar{3}$m*, *a* = 13.6Å

Coordination sequences and vertex symbols:

T$_1$ (32,3*m*)	4	9	19	34	48	66	96	127	151	183	4·6·4·6·4·6
T$_2$ (8, $\bar{4}$3*m*)	4	12	18	28	52	78	88	112	162	204	6·6·6·6·6·6

Secondary building units: 4-1

Composite building units:
d4r

Materials with this framework type:
*AlPO-16[1]
Octadecasil[2]
[Si$_n$Ge$_{40-n}$O$_{80}$]-**AST**, $0 \leq n \geq 40$[3]

AlPO-16   Type Material Data   **AST**

Crystal chemical data: $|(C_7H_{13}N)_4 (H_2O)_{16}| [Al_{20}P_{20}O_{80}]$-**AST**
                                  $C_7H_{13}N$ = quinuclidine
                                  cubic, $F23$, $a = 13.383$Å [1]

Framework density: 16.7 T/1000Å^3

Channels: apertures formed by 6-rings only

References:
(1) Bennett, J.M. and Kirchner, R.M. *Zeolites*, **11**, 502–506 (1991)
(2) Caullet, P., Guth, J.L., Hazm, J., Lamblin, J.M. and Gies, H. *Eur. J. Solid State Inorg. Chem.*, **28**, 345–361 (1991)
(3) Wang, Y.X., Song, J.Q. and Gies, H. *Solid State Sci.*, **5**, 1421–1433 (2003)

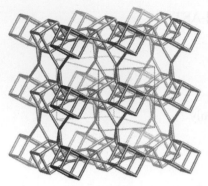 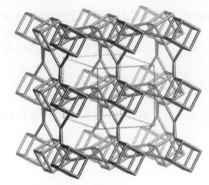

framework viewed along [001]

Idealized cell data: tetragonal, *P4/mcc*, *a* = 8.7Å, *c* = 13.9Å

Coordination sequences and vertex symbols:

T_1 (16,1)	4	9	19	35	52	72	100	131	163	201	$4 \cdot 6 \cdot 4 \cdot 6 \cdot 4 \cdot 6$
T_2 (4,222)	4	12	18	26	52	84	100	118	162	210	$6 \cdot 6 \cdot 6_2 \cdot 6_2 \cdot 12_8 \cdot 12_8$

Secondary building units: 4-1

Composite building units:

d4r   lau

Materials with this framework type:
*ASU-7[1]

Crystal chemical data: $|(C_2H_7N)_2 (H_2O)_2| [Ge_{20}O_{40}]$-**ASV**
       C_2H_7N = dimethylamine
   tetragonal, *P4/mcc*, a = 8.780Å, c = 14.470Å [1]

Framework density: 17.9 T/1000Å^3

Channels: [001] **12** 4.1x 4.1*

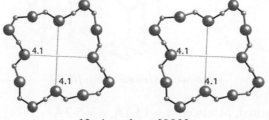

12-ring along [001]

References:
(1)   Li, H. and Yaghi, O.M. *J. Am. Chem. Soc.*, **120**, 10569–10570 (1998)

ATN

Framework Type Data

I4/mmm

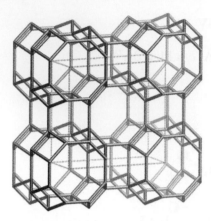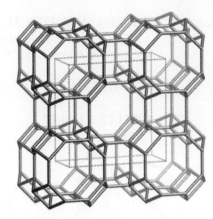

framework viewed along [001]

Idealized cell data: tetragonal, *I4/mmm*, *a* = 13.1Å, *c* = 5.3Å

Coordination sequences and vertex symbols:

| T$_1$ (16,*m*) | 4 | 10 | 21 | 36 | 54 | 78 | 106 | 136 | 173 | 214 | | 4·6·4·6·6·8 |

Secondary building units: 8 or 4

Composite building units:

        dzc   *atn*

double zigzag chain

Materials with this framework type:
  *MAPO-39[1,2]
  [Mg-Si-Al-P-O]-**ATN**[3]
  SAPO-39[4]
  ZnAPO-39[5]

MAPO-39 **Type Material Data** **ATN**

Crystal chemical data: $|H_n|$ $[Mg_nAl_{8-n}P_8O_{32}]$-**ATN**
tetragonal, $I4/m$, $a = 13.209$Å, $c = 5.277$Å [2]

Framework density: 17.4 T/1000Å^3

Channels: [001] **8** 4.0 x 4.0*

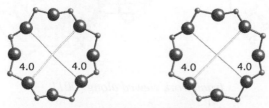

8-ring viewed along [001]

References:
(1) McCusker, L.B., Brunner, G.O. and Ojo, A.F. *Acta Crystallogr.*, **A46**, C59 (1990)
(2) Baur, W.H., Joswig, W., Kassner, D., Bieniok, A., Finger, G. and Kornatowski, J. *Z. Kristallogr.*, **214**, 154–159 (1999)
(3) Akporiaye, D.E., Andersen, A., Dahl, I.M., Mostad, H.B. and Wendelbo, R. *J. Phys. Chem.*, **99**, 14142–14148 (1995)
(4) Sinha, A.K., Hegde, S.G., Jacob, N.E. and Sivasanker, S. *Zeolites*, **18**, 350–355 (1997)
(5) Christensen, A.N., Jensen, T.R., Norby, P. and Hanson, J.C. *Chem. Mater.*, **10**, 1688–1693 (1998)

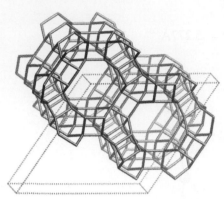 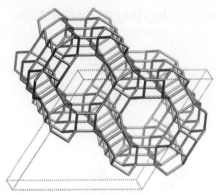 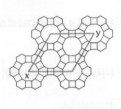

framework viewed along [001]

Idealized cell data: trigonal, $R\overline{3}m$, $a = 20.9$Å, $c = 5.1$Å

Coordination sequences and vertex symbols:

T₁ (36,1)   4   11   22   37   59   85   114   147   184   230 $4 \cdot 6_2 \cdot 6 \cdot 6_2 \cdot 6 \cdot 6_3$

Secondary building units: 12 or 6 or 4

Composite building units:
lau

Materials with this framework type:
    *AlPO-31[1,2]
    MAPO-31, M = Mn, Ni, Zn[3]
    MAPO-31, M = Zn, Mg, Mn, Co, Cr, Cu, Cd[4]

    SAPO-31[5-7]
    VAPO-31[8]

AlPO-31   **Type Material Data**   # ATO

Crystal chemical data: $[Al_{18}P_{18}O_{72}]$-**ATO**

trigonal, $R\bar{3}$, $a = 20.827$Å, $c = 5.003$Å [1]

Framework density: 19.2 T/1000Å³

Channels: [001] **12** 5.4 x 5.4*

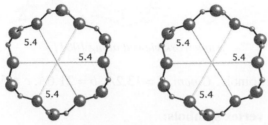

12-ring viewed along [001]

References:

(1) Bennett, J.M. and Kirchner, R.M. *Zeolites*, **12**, 338–342 (1992)

(2) Baur, W.H., Joswig, W., Kassner, D. and Kornatowski, J. *Acta Crystallogr.*, **B50**, 290–294 (1994)

(3) Umamaheswari, V., Kanna, C., Arabindoo, B., Palanichamy, M. and Murugesan, V. *Proc. Indian Acad. Sci. (Chem. Sci.)*, **112**, 439–448 (2000)

(4) Finger, G., Kornatowski, J., Jancke, K., Matschat, R. and Baur, W.H. *Microporous Mesoporous Mat.*, **33**, 127–136 (1999)

(5) Flanigen, E.M., Lok, B.M., Patton, R.L. and Wilson, S.T. *Pure Appl. Chem.*, **58**, 1351–1358 (1986)

(6) Flanigen, E.M., Lok, B.M., Patton, R.L. and Wilson, S.T. *Proc. 7th Int. Zeolite Conf.*, pp. 103–112 (1986)

(7) Baur, W.H., Joswig, W., Kassner, D. and Kornatowski, J. *Acta Crystallogr.*, **B50**, 290–294 (1994)

(8) Venkatathri, N., Gegde, S.G. and Sivasanker, S. *Chem. Commun.*, 151–152 (1995)

ATS

Framework Type Data

Cmcm

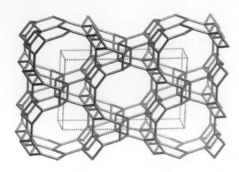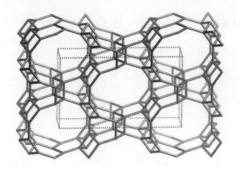

framework viewed along [001]

Idealized cell data: orthorhombic, *Cmcm*, $a = 13.2$Å, $b = 21.6$Å, $c = 5.3$Å

Coordination sequences and vertex symbols:

T_1 (8,*m*)	4	10	19	30	46	67	93	124	154	189	$4·6·4·6·6·6_2$
T_2 (8,*m*)	4	10	20	32	49	73	97	124	157	193	$4·6_2·4·6_2·6·12_2$
T_3 (8,*m*)	4	10	19	32	51	72	96	124	155	196	$4·6_2·4·6_2·6·12_2$

Secondary building units: 12 or 6 or 4

Composite building units:

       dzc   *ats*

    double zigzag
        chain

Materials with this framework type:
    *MAPO-36[1]
    AlPO-36[2]
    FAPO-36[3]

    SSZ-55[4]
    ZnAPO-36[5]

MAPO-36   **Type Material Data**   **ATS**

Crystal chemical data: $|H|$ $[MgAl_{11}P_{12}O_{48}]$-**ATS**
monoclinic, $C2/c$

$$a = 13.148\text{Å},\ b = 21.577\text{Å},\ c = 5.164\text{Å},\ \beta = 91.84°\,^{(1)}$$

Framework density: 16.4 T/1000Å^3

Channels: [001] **12** 6.5 x 7.5*

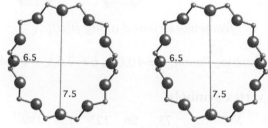

12-ring viewed along [001]

References:
(1) Smith, J.V., Pluth, J.J. and Andries, K.J. *Zeolites*, **13**, 166–169 (1993)
(2) Zahedi-Niaki, M.H., Xu, G.Y., Meyer, H., Fyfe, C.A. and Kaliaguine, S. *Microporous Mesoporous Mat.*, **32**, 241–250 (1999)
(3) Ristic, A., Tusar, N.N., Arcon, I., Logar, N.Z., Thibault-Starzyk, F., Czyzniewska, J. and Kaucic, V. *Chem. Mater.*, **15**, 3643–3649 (2003)
(4) Wu, M.G., Deem, M.W., Elomari, S.A., Medrud, R.C., Zones, S.I., Maesen, T., Kibby, C., Chen, C.-Y. and Chen, I.Y. *J. Phys. Chem. B*, **106**, 264–270 (2002)
(5) Christensen, A.N., Norby, P. and Hanson, J.C. *Microporous Mesoporous Mat.*, **20**, 349–354 (1998)

ATT

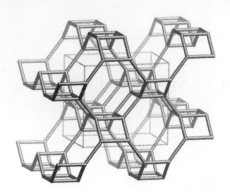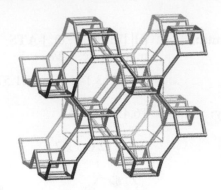

framework viewed along [010]

Idealized cell data: orthorhombic, *Pmma*, a = 10.0Å, b = 7.5Å, c = 9.4Å

Coordination sequences and vertex symbols:

T_1 (8,1)	4	9	18	33	52	73	96	123	158	199	4·4·4·6·8·8
T_2 (4,m)	4	10	21	34	48	70	100	130	159	194	$4·8_2·4·8_2·6·8_2$

Secondary building units: 4-2 or 6

Composite building units:

dcc	*dsc*	*abw*	*gis*
double crankshaft chain	double sawtooth chain		

Materials with this framework type:
*AlPO-12-TAMU[1]
AlPO-33[2]
AlPO-33[3]
RMA-3[4]

AlPO-12-TAMU **Type Material Data** **ATT**

Crystal chemical data: |(C₄H₁₂N)₄ (OH)₄| [Al₁₂P₁₂O₄₈]-**ATT**

$$|(C_4H_{12}N)_4\,(OH)_4|\,[Al_{12}P_{12}O_{48}]\text{-}\textbf{ATT}$$

C₄H₁₂N = tetramethylammonium
orthorhombic, $P2_12_12$, $a = 10.332$Å, $b = 14.640$Å, $c = 9.511$Å [1]
(Relationship to unit cell of Framework Type: $a' = a$, $b' = 2b$, $c' = c$)

Framework density: 16.7 T/1000Å³

Channels: [100] **8** 4.2 x 4.6* ↔ [010] **8** 3.8 x 3.8*

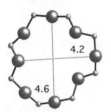 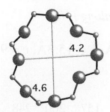

8-ring viewed along [100]

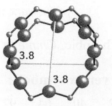 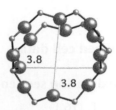

complex 8-ring viewed along [010]

References:
(1) Rudolf, P.R., Saldarriaga-Molina, C. and Clearfield, A. *J. Phys. Chem.*, **90**, 6122–6125 (1986)
(2) Smith, J.V., Pluth, J.J. and Bennett, J.M. *private communication*
(3) Patton, R.L. and Gajek, R.T. *U.S. Patent 4,473,663* (1984)
(4) Ikeda, T. and Itabashi, K. *Chem. Commun.*, 2753–2755 (2005)

ATV

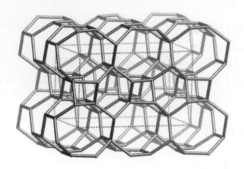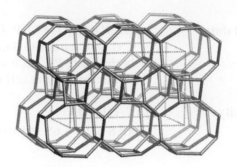

framework viewed along [100]

Idealized cell data: orthorhombic, *Cmme*, $a = 8.6$Å, $b = 15.3$Å, $c = 9.7$Å

Coordination sequences and vertex symbols:

T_1 (16,1)	4	11	22	40	64	92	121	157	200	248	$4 \cdot 6_2 \cdot 6 \cdot 6_3 \cdot 6_2 \cdot 6_3$
T_2 (8,m)	4	12	25	42	61	88	122	160	200	246	$6 \cdot 6_2 \cdot 6 \cdot 6_2 \cdot 6_2 \cdot 6_2$

Secondary building units: 6 or 4-[1,1]

Composite building units:

      nsc   *afi*   *bog*

   narsarsukite
      chain

Materials with this framework type:
    *AlPO-25[1]
    [Ga-P-O]-**ATV**[2]

AlPO-25 **Type Material Data** **ATV**

Crystal chemical data: [Al$_{12}$P$_{12}$O$_{48}$]-**ATV**
orthorhombic, *Aemm*, $a = 9.449$Å, $b = 15.203$Å, $c = 8.408$Å [1]
(Relationship to unit cell of Framework Type: $a' = c$, $b' = b$, $c' = a$)

Framework density: 19.9 T/1000Å^3

Channels: [001] **8** 3.0 x 4.9*

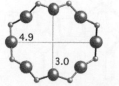 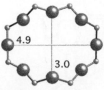

8-ring viewed along [001]

References:
(1) Richardson Jr., J.W., Smith, J.V. and Pluth, J.J. *J. Phys. Chem.*, **94**, 3365–3367 (1990)
(2) Parise, J.B. *Chem. Commun.*, 606–607 (1985)

AWO

Framework Type Data

Cmce

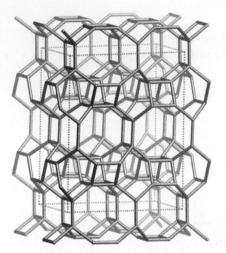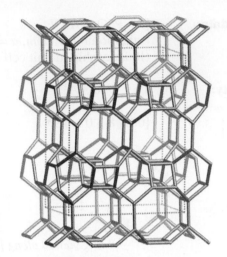

framework viewed along [100]

Idealized cell data: orthorhombic, *Cmce*, *a* = 9.1Å, *b* = 15.0Å, *c* = 19.2Å

Coordination sequences and vertex symbols:

T_1 (16,1)	4	10	20	35	55	78	103	133	173	217		$4 \cdot 6_2 \cdot 4 \cdot 8_3 \cdot 6 \cdot 8_2$
T_2 (16,1)	4	10	21	36	53	76	108	142	173	210		$4 \cdot 6 \cdot 4 \cdot 8_2 \cdot 6 \cdot 8$
T_3 (16,1)	4	9	19	35	54	76	102	134	172	214		$4 \cdot 4 \cdot 4 \cdot 6 \cdot 8 \cdot 8_3$

Secondary building units: 6 or 4-2 or 4

Composite building units:
dcc
double
crankshaft chain

Materials with this framework type:
*AlPO-21[1,2]
[Ga-P-O]-**AWO**[3]

AlPO-21   **Type Material Data**   **AWO**

Crystal chemical data: $|H_4 (C_2H_7N)_{10.66} (C_3H_8)_{5.33} (OH)_4|$ $[Al_{12}P_{12}O_{48}]$-**AWO**
C_2H_7N = dimethylamine, C_3H_8 = propane
monoclinic, $P2_1/a$

$a = 10.330$Å, $b = 17.524$Å, $c = 8.676$Å, $\beta = 123.37°$ [1]
(Relationship to unit cell of Framework Type:

$a' = a$, $b' = c$, $c' = b/(2\sin\beta')$
or, as vectors, $\mathbf{a}' = \mathbf{a}$, $\mathbf{b}' = \mathbf{c}$, $\mathbf{c}' = (\mathbf{b} - \mathbf{a})/2$)

Stability: Transforms to AlPO-25 (**ATV**) upon calcination [2]

Framework density: 18.3 T/1000Å^3

Channels: [100] **8** 2.7 x 5.5*

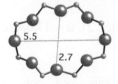 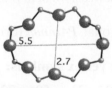

8-ring viewed along [100]

References:
(1) Bennett, J.M., Cohen, J.M., Artioli, G., Pluth, J.J. and Smith, J.V. *Inorg. Chem.*, **24**, 188–193 (1985)
(2) Parise, J.B. and Day, C.S. *Acta Crystallogr.*, **C41**, 515–520 (1985)
(3) Parise, J.B. *Chem. Commun.*, 606–607 (1985)

AWW

Framework Type Data

P4/nmm

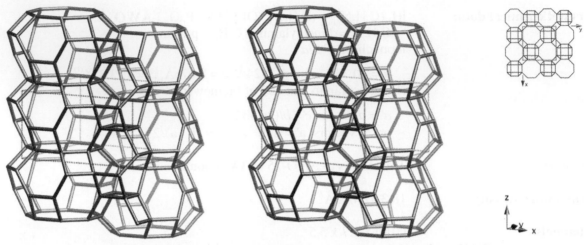

framework viewed normal to [001] (upper right: projection down [001])

Idealized cell data: tetragonal, *P4/nmm* (origin choice 2), $a = 13.6$Å, $c = 7.6$Å

Coordination sequences and vertex symbols:

T$_1$ (16,1)	4	10	20	33	50	72	98	128	162	200	4·4·6·6·6·8
T$_2$ (8,2)	4	9	17	30	50	74	97	123	158	198	4·4·4·6·6·6

Secondary building units: 6 or 4

Composite building units:

       aww   *clo*

Materials with this framework type:
    *AlPO-22[1]
    AlPO-CJB1 (additional phosphate group present)[2]

AIPO-22   **Type Material Data**   **AWW**

Crystal chemical data: $|(C_7H_{14}N)_4 \ (HPO_4)_2| \ [Al_{24}P_{24}O_{96}]$-**AWW**
       $C_7H_{14}N$ = quinuclidinium
       tetragonal, $P4/ncc$, a = 13.628Å, c = 15.463Å [1]
       (Relationship to unit cell of Framework Type: $a'= a$, $c' = 2c$)

Framework density: 16.7 T/1000Å^3

Channels: [001] **8** 3.9 x 3.9*

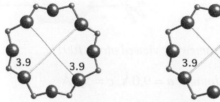

8-ring viewed along [001]

References:
(1) Richardson Jr., J.W., Pluth, J.J. and Smith, J.V. *Naturwiss.*, **76**, 467–469 (1989)
(2) Yan, W.F., Yu, J.H., Xu, R.R., Zhu, G.S, Xiao, F.S., Han, Y., Sugiyama, K. and Terasaki, O. *Chem. Mater.*, **12**, 2517–2519 (2000)

BCT

Framework Type Data

I4/mmm

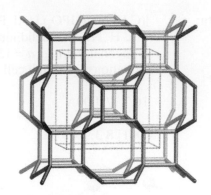

framework viewed along [001]

Idealized cell data: tetragonal, *I4/mmm*, $a = 9.0$Å, $c = 5.3$Å

Coordination sequences and vertex symbols:

| $T_1(8,m.2m)$ | 4 | 11 | 24 | 41 | 62 | 90 | 122 | 157 | 200 | 247 | | $4·6_2·6·6·6·6$ |

Secondary building units: 8 or 4

Composite building units:
lau

Materials with this framework type:

*Mg-BCTT[1]

|Ca-|[Al-Si-O]-**BCT**[2]

Fe(III)-BCTT[1]

Metavariscite[3]

Svyatoslavite[4]

Zn-BCTT[1]

Mg-BCTT **Type Material Data** **BCT**

Crystal chemical data: |K$_{4.56}$|[Mg$_{2.28}$Si$_{5.72}$O$_{16}$]-**BCT**
tetragonal, *I4mm*, $a = 8.957$Å, $c = 5.281$Å [1]

Framework density: 18.9 T/1000Å^3

Channels: [001] **8** 2.4 x 2.4*

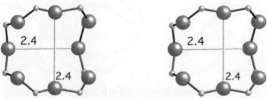

8-ring viewd along [001]

References:
(1) Dollase, W.A. and Ross, C.R. *Am. Mineral.*, **78**, 627–632 (1993)
(2) Takeuchi, Y., Haga, N. and Ito, J. *Z. Kristallogr.*, **137**, 380–398 (1973)
(3) Kniep, R. and Mootz, D. *Acta Crystallogr.*, **B29**, 2292–2294 (1973)
(4) Chesnokov, B.V., Lotova, E.V., Pavlyuchenko, V.S., Usova, L.V., Bushmakin, A.F. and Nishanbayev, T.P. *Zap. Vses. Mineral. Obshch. (Am. Mineral. 76, 299–301 (1991))*, **118**, 111–114 (1989)

*BEA

Framework Type Data

*P*4₁22

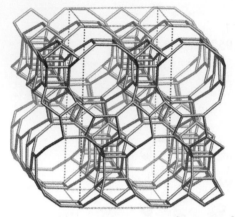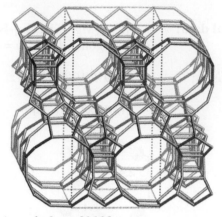

framework viewed along [010]

Idealized cell data: tetragonal, $P4_122$, $a = 12.6$Å, $c = 26.2$Å

Coordination sequences and vertex symbols:

T_1 (8,1)	4	10	19	32	51	77	105	133	167	207	$4·5·4·12_3·5·5$
T_2 (8,1)	4	10	19	32	51	75	102	133	170	208	$4·5·4·12_6·5·5$
T_3 (8,1)	4	10	21	32	49	76	109	137	170	207	$4·6·4·12_3·5·5$
T_4 (8,1)	4	10	21	32	49	74	105	139	173	204	$4·6·4·12_6·5·5$
T_5 (8,1)	4	11	18	29	48	80	107	133	160	203	$4·5_2·5·5·5·6$
T_6 (8,1)	4	11	18	29	48	77	106	134	160	204	$4·5_2·5·5·5·6$
T_7 (8,1)	4	12	18	31	51	76	109	133	164	210	$5·5·5·6·5_2·12_5$
T_8 (4,2)	4	12	19	32	48	75	112	134	164	206	$5·5·5_2·12_7·6·6$
T_9 (4,2)	4	12	17	30	54	77	106	134	160	212	$5·5·5·5·5_2·12_3$

Secondary building units: see *Compendium*

Composite building units:

mor	bea	mtw

Materials with this framework type:

*Beta[1,2]
[B-Si-O]-*BEA[3,4]
[Ga-Si-O]-*BEA[4]
[Ti-Si-O]-*BEA[5]

Al-rich beta[6]
CIT-6[7]
Tschernichite[8]
pure silica beta[9]

Beta polymorph A   Type Material Data   ***BEA**

Crystal chemical data: |Na$_7$| [Al$_7$Si$_{57}$O$_{128}$]-***BEA**
tetragonal, $P4_122$, a = 12.661Å, c = 26.406Å [2]

Framework density: 15.1 T/1000Å^3

Channels: <100> **12** 6.6 x 6.7** ↔ [001] **12** 5.6 x 5.6*

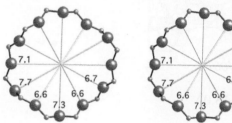
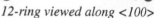

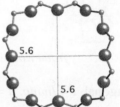

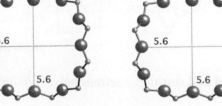

12-ring viewed along <100>   *12-ring viewed along [001]*

References:
(1) Higgins, J.B., LaPierre, R.B., Schlenker, J.L., Rohrman, A.C., Wood, J.D., Kerr, G.T. and Rohrbaugh, W.J. *Zeolites*, **8**, 446–452 (1988)
(2) Newsam, J.M., Treacy, M.M.J., Koetsier, W.T. and de Gruyter, C.B. *Proc. R. Soc. Lond. A*, **420**, 375–405 (1988)
(3) Marler, B., Böhme, R. and Gies, H. *Proc. 9th Int. Zeolite Conf.*, pp. 425–432 (1993)
(4) Reddy, K.S.N., Eapen, M.J., Joshi, P.N., Mirajkar, S.P. and Shiralkar, V.P. *J. Incl. Phenom. Mol. Recogn. Chem.*, **20**, 197–210 (1994)
(5) Blasco, T., Camblor, M.A., Corma, A., Esteve, P., Martinez, A., Prieto, C. and Valencia, S. *Chem. Commun.*, 2367–2368 (1996)
(6) Borade, R.B. and Clearfield, A. *Microporous Materials*, **5**, 289–297 (1996)
(7) Takewaki, T., Beck, L.W. and Davis, M.E. *Topics in Catalysis*, **9**, 35–42 (1999)
(8) Boggs, R.C., Howard, D.G., Smith, J.V. and Klein, G.L. *Am. Mineral.*, **78**, 822–826 (1993)
(9) Camblor, M.A., Corma, A. and Valencia, S. *Chem. Commun.*, 2365–2366 (1996)

BEC

Framework Type Data

P4₂/mmc

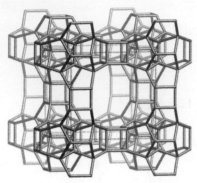
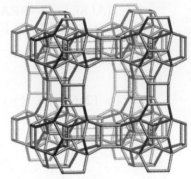

framework viewed along [001]

Idealized cell data: tetragonal, *P4₂/mmc*, *a* = 12.8Å, *c* = 13.0Å

Coordination sequences and vertex symbols:

T₁(16,1)	4	9	18	32	50	71	96	129	167	199	4·5·4·6·4·12₆
T₂(8,2)	4	12	17	30	48	71	98	126	156	198	5·5·5₂·12₅·6·6
T₃(8,*m*)	4	11	20	28	41	70	103	127	150	188	4·5₂·5·6·5·6

Secondary building units: 6-2

Composite building units:

       d4r   *mor*   *mtw*

Materials with this framework type:
*FOS-5 (Beta polymorph C)[1]
ITQ-14 overgrowth[2]
ITQ-17[3]

(Beta polymorph C)   **Type Material Data**   **BEC**

Crystal chemical data: $|(C_3H_9N)_{48}\ (H_2O)_{36}|\ [Ge_{256}O_{512}]$-**BEC**
tetragonal, $I4_1/amd$, $a = 25.990$Å, $c = 27.271$Å $^{(1)}$
(Relationship to unit cell of Framework Type: $a' = 2a$, $c' = 2c$)

Framework density: 13.9 T/1000Å^3

Channels: [001] **12** 6.3 x 7.5* $\leftrightarrow$ <100> **12** 6.0 x 6.9**

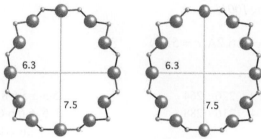

12-ring viewed along [001]

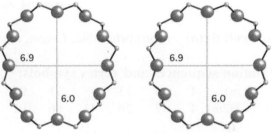

12-ring viewed along [100]

References:
(1) Conradsson, T., Dadachov, M.S. and Zou, X.D. *Microporous Mesoporous Mat.*, **41**, 183–191 (2000)
(2) Liu, Z., Ohsuna, T., Terasaki, O., Camblor, M.A., Diaz-Cabañas, M.-J. and Hiraga, K. *J. Am. Chem. Soc.*, **123**, 5370–5371 (2001)
(3) Corma, A., Navarro, M.T., Rey, F., Rius, J. and Valencia, S. *Angew. Chem., Int. Ed.*, **40**, 2277–2280 (2001)

BIK

Framework Type Data

Cmcm

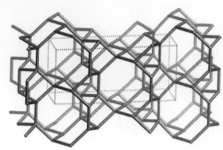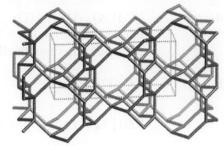

framework viewed along [001]

Idealized cell data: orthorhombic, *Cmcm*, a = 7.5Å, b = 16.2Å, c = 5.3Å

Coordination sequences and vertex symbols:

T_1 (8,*m*)	4	12	23	43	71	97	128	179	226	264
T_2 (4,*m2m*)	4	12	26	42	66	102	140	164	216	288

$5 \cdot 5 \cdot 5 \cdot 5 \cdot 6 \cdot 8_2$

$5_2 \cdot 6_2 \cdot 6 \cdot 6 \cdot 6 \cdot 6$

Secondary building units: 5-1

Composite building units:
bik

Materials with this framework type:
   *Bikitaite[1,2]
   |Cs-|[Al-Si-O]-**BIK**[3]
   Triclinic bikitaite[4]

Crystal chemical data:

|Li$_2$(H$_2$O)$_2$| [Al$_2$Si$_4$O$_{12}$]-**BIK**
triclinic, $P1$, $a = 8.607$Å, $b = 4.954$Å, $c = 7.597$Å

$\alpha = 89.90°$, $\beta = 114.44°$, $\gamma = 89.99°$ [2]
(Relationship to unit cell of Framework Type:

$a' = b/(2\sin\beta')$, $b' = c$, $c' = a$
or, as vectors, $\mathbf{a'} = (\mathbf{b} - \mathbf{a})/2$, $\mathbf{b'} = \mathbf{c}$, $\mathbf{c'} = \mathbf{a}$)

Framework density: 20.3 T/1000Å^3

Channels: [010] **8** 2.8 x 3.7*

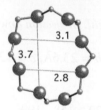 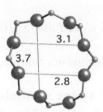

8-ring viewed along [010]

References:
(1) Kocman, V., Gait, R.I. and Rucklidge, J. *Am. Mineral.*, **59**, 71–78 (1974)
(2) Ståhl, K., Kvick, Å. and Ghose, S. *Zeolites*, **9**, 303–311 (1989)
(3) Annehed, H. and Fälth, L. *Z. Kristallogr.*, **166**, 301–306 (1984)
(4) Bissert, G. and Liebau, L. *N. Jb. Miner. Mh.*, 241–252 (1986)

BOG

Framework Type Data

Imma

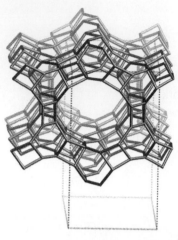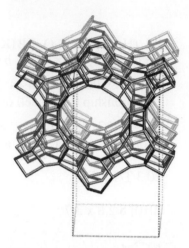

framework viewed along [100] (upper right: projection down [100])

Idealized cell data: orthorhombic, *Imma*, $a = 20.0$Å, $b = 23.6$Å, $c = 12.7$Å

Coordination sequences and vertex symbols:

T_1 (16,1)	4	10	19	32	51	74	101	129	158	199	$4·5·4·6·5·12_2$
T_2 (16,1)	4	10	20	32	48	74	104	131	159	195	$4·5·4·6_2·10·12$
T_3 (16,1)	4	11	19	34	50	71	98	133	162	195	$4·5_2·5·6·5·10_3$
T_4 (16,1)	4	11	21	32	49	74	101	128	162	200	$4·10_5·5·6_3·5·6_3$
T_5 (16,1)	4	11	20	31	53	76	97	126	168	199	$4·5·5·6_2·5·10$
T_6 (16,1)	4	11	18	31	52	75	100	126	158	206	$4·5·5·6·5·6_2$

Secondary building units: 6 or 5-1 or 4

Composite building units:

bre   bog   cas

Materials with this framework type:
   *Boggsite[1]
   Dehyd. boggsite[2]

Boggsite   **Type Material Data**   **BOG**

Crystal chemical data: $|Ca_7Na_4(H_2O)_{74}|$ $[Al_{18}Si_{78}O_{192}]$-**BOG**
orthorhombic, *Imma*, $a = 20.236$Å, $b = 23.798$Å, $c = 12.798$Å [1]

Framework density: 15.6 T/1000Å^3

Channels: [100] **12** 7.0 x 7.0* ↔ [010] **10** 5.5 x 5.8*

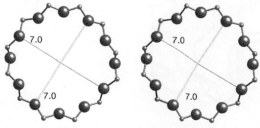 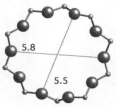

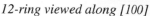

12-ring viewed along [100]   *10-ring viewed along [010]*

References:
(1) Pluth, J.J. and Smith, J.V. *Am. Mineral.*, **75**, 501–507 (1990)
(2) Zanardi, S., Cruciani, G., Alberti, A. and Galli, E. *Am. Mineral.*, **89**, 1033–1042 (2004)

BPH

Framework Type Data

$P\bar{6}2m$

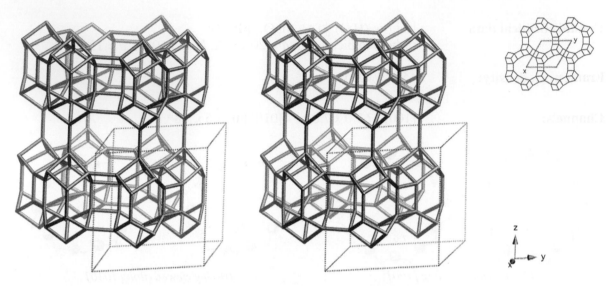

framework viewed normal to [001] (upper right: projection down [001])

Idealized cell data: hexagonal, $P\bar{6}2m$, a = 13.1Å, c = 13.0Å

Coordination sequences and vertex symbols:

T_1 (12,1)	4	9	17	28	42	60	84	113	140	169	$4 \cdot 4 \cdot 4 \cdot 8_2 \cdot 6_2 \cdot 8$
T_2 (12,1)	4	9	16	25	39	61	86	111	141	173	$4 \cdot 4 \cdot 4 \cdot 6 \cdot 6 \cdot 12$
T_3 (4,3)	4	9	18	30	43	62	85	105	135	180	$4 \cdot 8 \cdot 4 \cdot 8 \cdot 4 \cdot 8$

Secondary building units: 6*1

Composite building units:

        afs   *bph*

Materials with this framework type:
    *Beryllophosphate-H[1,2]
    Linde Q[3]
    STA-5[4]
    UZM-4[5]

Beryllophosphate-H   Type Material Data   **BPH**

Crystal chemical data: $|K_7Na_7(H_2O)_{20}|\ [Be_{14}P_{14}O_{56}]$-**BPH**
trigonal, $P321$, $a = 12.582Å$, $c = 12.451Å$ [(2)]

Framework density: $16.4\ T/1000Å^3$

Channels: [001] **12** 6.3 x 6.3* ↔ ⊥ [001] **8** 2.7 x 3.5**

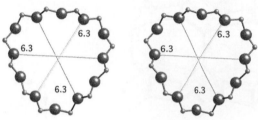 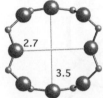 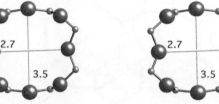

12-ring viewed along [001]   *8-ring viewed normal to [001]*

References:
(1) Harvey, G. *Z. Kristallogr.*, **182**, 123–124 (1988)
(2) Harvey, G., Baerlocher, Ch. and Wroblewski, T. *Z. Kristallogr.*, **201**, 113–123 (1992)
(3) Andries, K.J., Bosmans, H.J. and Grobet, P.J. *Zeolites*, **11**, 124–131 (1991)
(4) Patinec, V., Wright, P.A., Aitken, R.A., Lightfoot, P., Purdie, S.D.J., Cox, P.A., Kvick, A. and Vaughan, G. *Chem. Mater.*, **11**, 2456–2462 (1999)
(5) Blackwell, C.S., Broach, R.W., Gatter, M.G., Holmgren, J.S., Jan, D.-Y., Lewis, G.J., Mezza, B.J., Mezza, T.M., Miller, M.A., Moscoso, J.G., Patton, R.L., Rohde, L.M., Schoonover, M.W., Sinkler, W., Wilson, B.A. and Wilson, S.T. *Angew. Chem., Int. Ed.*, **42**, 1737–1740 (2003)

BRE

Framework Type Data

P2₁/m

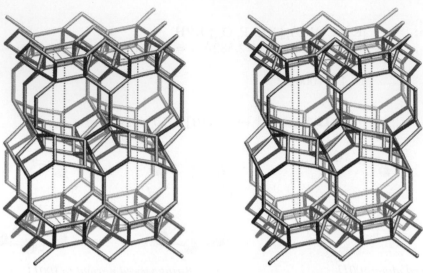

framework viewed along [100]

Idealized cell data: monoclinic, $P2_1/m$, $a = 6.8$Å, $b = 17.1$Å, $c = 7.6$Å, $\beta = 95.8°$ [2]

Coordination sequences and vertex symbols:

T₁ (4,1)	4	10	20	37	61	83	110	144	192	238	4·5·4·6·5·8
T₂ (4,1)	4	10	20	36	61	85	107	147	191	234	4·5·4·8·5·6
T₃ (4,1)	4	11	23	37	54	82	119	152	184	233	4·8·5·8·5·8₂
T₄ (4,1)	4	11	18	37	62	85	110	147	195	236	4·5₂·5·6·5·8

Secondary building units: 4

Composite building units:
    bre

Materials with this framework type:
    *Brewsterite[1,2]
    Ba-dominant brewsterite[3]
    CIT-4[4]

    Dehyd. brewsterite[5]
    Synthetic brewsterite[6]

Brewsterite   Type Material Data   **BRE**

Crystal chemical data: $|(Ba,Sr)_2 (H_2O)_{10}| [Al_4Si_{12}O_{32}]$-**BRE**
monoclinic, $P2_1/m$

$$a = 6.793\text{Å}, b = 17.573\text{Å}, c = 7.759\text{Å}, \beta = 94.54°^{(2)}$$

Framework density: 17.3 T/1000Å^3

Channels: [100] **8** 2.3 x 5.0* ↔ [001] **8** 2.8 x 4.1*

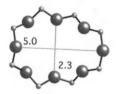 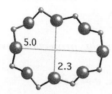 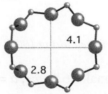 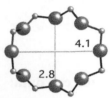

8-ring viewed along [100]   *8-ring viewed along [001]*

References:
(1) Perrotta, A.J. and Smith, J.V. *Acta Crystallogr.*, **17**, 857–862 (1964)
(2) Schlenker, J.L., Pluth, J.J. and Smith, J.V. *Acta Crystallogr.*, **B33**, 2907–2910 (1977)
(3) Cabella, R., Lucchetti, G., Palenzona, A., Quartieri, S. and Vezzalini, G. *Eur. J. Mineral.*, **5**, 353–360 (1993)
(4) Khodabandeh, S., Lee, G. and Davis, M.E. *Microporous Mesoporous Mat.*, **11**, 87–95 (1997)
(5) Ståhl, K. and Hanson, J.C. *Microporous Mesoporous Mat.*, **32**, 147–158 (1999)
(6) Ghobarkar, H. and Schaef, O. *German Patent AZ 198 24 184.4–41* (1999)

CAN

Framework Type Data

$P6_3/mmc$

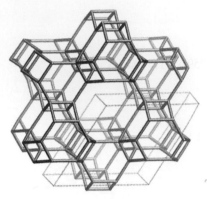 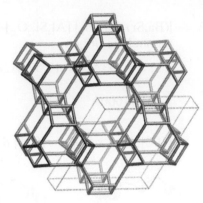 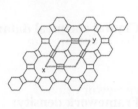

framework viewed along [001] (upper right: projection down [001])

Idealized cell data: hexagonal, $P6_3/mmc$, $a = 12.5$Å, $c = 5.3$Å

Coordination sequences and vertex symbols:

T_1 (12,*m*)	4	10	20	34	54	78	104	134	168	210	4·6·4·6·6·6

Secondary building units: 12 or 6 or 4

Framework description: AB sequence of 6-rings

Composite building units:

        dzc   *can*

      double zigzag
         chain

Materials with this framework type:

*Cancrinite[1,2]
[Al-Ge-O]-**CAN**[3]
[Co-P-O]-**CAN**[4]
[Ga-Ge-O]-**CAN**[5]
[Ga-Si-O]-**CAN**[6]
[Zn-P-O]-**CAN**[7]

|Li-Cs|[Al-Si-O]-**CAN**[8]
|Li-Tl|[Al-Si-O]-**CAN**[8]
Basic cancrinite[9,10]
Cancrinite hydrate[11]
Davyne[12]
ECR-5[13]

Microsommite[14]
Synthetic cancrinite[15]
Tiptopite[16]
Vishnevite[17]

Cancrinite   Type Material Data   **CAN**

Crystal chemical data: $|Na_6Ca(H_2O)_2CO_3|[Al_6Si_6O_{24}]$-**CAN**
hexagonal, $P6_3$, $a = 12.75$Å, $c = 5.14$Å [(2)]

Framework density: 16.6 T/1000Å^3

Channels: [001] **12** 5.9 x 5.9*

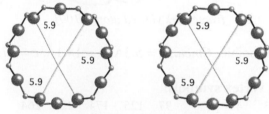

12-ring viewed along [001]

References:
(1) Pauling, L. *Proc. Natl. Acad. Sci.*, **16**, 453–459 (1930)
(2) Jarchow, O. *Z. Kristallogr.*, **122**, 407–422 (1965)
(3) Belokoneva, E.L., Uvarova, T.G. and Dem'yanets, L.N. *Sov. Phys. Crystallogr.*, **31**, 516–519 (1986)
(4) Bieniok, A., Brendel, U., Paulus, E.F. and Amthauer, G. *Eur. J. Mineral.*, **17**, 813–818 (2005)
(5) Lee, Y., Parise, J.B., Tripathi, A., Kim, S.J. and Vogt, T. *Microporous Mesoporous Mat.*, **39**, 445–455 (2000)
(6) Newsam, J.M. and Jorgensen, J.D. *Zeolites*, **7**, 569-573 (1987)
(7) Yakubovich, O.V., Karimova, O.V. and Mel'nikov, O.K. *Crystallogr. Reports*, **39**, 564-568 (1994)
(8) Norby, P., Krogh Andersen, I.G., Krogh Andersen, E., Colella, C. and de'Gennaro, M. *Zeolites*, **11**, 248-253 (1991)
(9) Barrer, R.M. and White, E.A.D. *J. Chem. Soc.*, 1561-1571 (1952)
(10) Bresciana Pahor, N., Calligaris, M., Nardin, G. and Randaccio, L. *Acta Crystallogr.*, **B38**, 893-895 (1982)
(11) Wyart, J. and Michel-Levy, M. *Compt. Rend.*, **229**, 131- (1949)
(12) Hassan, I. and Grundy, H.D. *Can. Mineral.*, **28**, 341-349 (1990)
(13) Vaughan, D.E.W. *E. Patent A-190,90* (1986)
(14) Bonaccorsi, E., Comodi, P. and Merlino, S. *Phys. Chem. Mineral.*, **22**, 367-374 (1995)
(15) Smolin, Y.I., Shepelev, Y.F., Butikova, I.K. and Kobyakov, I.B. *Kristallografiya*, **26**, 63-66 (1981)
(16) Peacor, D.R., Rouse, R.C. and Ahn, J.-H. *Am. Mineral.*, **72**, 816-820 (1987)
(17) Hassan, I. and Grundy, H.D. *Can. Mineral.*, **22**, 333-340 (1984)

CAS

Framework Type Data

Cmcm

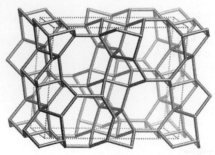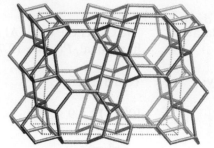

framework viewed along [100]

Idealized cell data: orthorhombic, *Cmcm*, $a = 5.3$Å, $b = 14.1$Å, $c = 17.2$Å

Coordination sequences and vertex symbols:

T_1 (8,*m*)	4	12	23	41	70	97	125	174	224	264	$5 \cdot 5 \cdot 5 \cdot 5 \cdot 6 \cdot 8_2$
T_2 (8,*m*)	4	12	26	43	64	101	138	165	215	284	$5 \cdot 6 \cdot 5 \cdot 6 \cdot 6_2 \cdot 8_2$
T_3 (8,*m*)	4	12	23	43	72	95	128	177	225	259	$5 \cdot 6 \cdot 5 \cdot 6 \cdot 5_2 \cdot 6$

Secondary building units: 5-1

Composite building units:

cas   *bik*

Materials with this framework type:
 *Cesium Aluminosilicate[1,2]
 EU-20b (**CAS-NSI** structural intermediate)[3]

Cesium Aluminosilicate   Type Material Data   **CAS**

Crystal chemical data: |Cs$_4$| [Al$_4$Si$_{20}$O$_{48}$]-**CAS**
orthorhombic, *Ama*2, $a = 16.776$Å, $b = 13.828$Å, $c = 5.021$Å [1]
(Relationship to unit cell of Framework Type: $a' = -c$, $b' = b$, $c' = a$)

Framework density: 20.6 T/1000Å^3

Channels: [001] **8** 2.4 x 4.7*

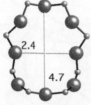 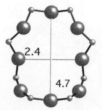

8-ring viewed along [001]

References:
(1) Araki, T. *Z. Kristallogr.*, **152**, 207–213 (1980)
(2) Hughes, R.W. and Weller, M.T. *Microporous Mesoporous Mat.*, **51**, 189–196 (2002)
(3) Marler, B., Camblor, M.A. and Gies, H. *Microporous Mesoporous Mat.*, **90**, 87–101 (2006)

CDO

Framework Type Data

Cmcm

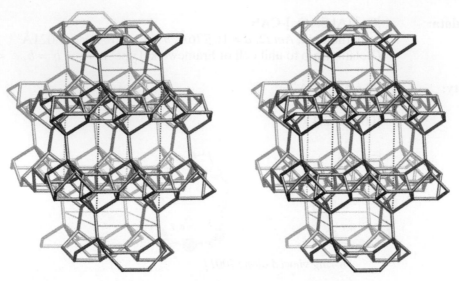

framework viewed along [001]

Idealized cell data: orthorhombic, *Cmcm*, a = 7.6Å, b = 18.7Å, c = 14.1Å

Coordination sequences and vertex symbols:

T$_1$ (16,1)	4	12	21	39	67	99	129	172	228	275	319	391	5·5·5·5$_2$·5·8
T$_2$ (8,..m)	4	12	20	35	69	105	125	168	231	282	320	381	5·5$_2$·5·5$_2$·8·*
T$_3$ (8,m..)	4	12	28	44	64	100	144	178	215	277	347	402	5·8$_2$·5·8$_2$·5$_2$·8$_2$
T$_4$ (4,$m2m$)	4	12	24	42	66	98	130	172	228	278	334	398	5·5·5·5·5$_2$·8$_2$

Secondary building units: 5-1

Composite building units:

fer

Materials with this framework type:
  *CDS-1[1]
  MCM-65[2]
  UZM-25[3]

CDS-1   **Type Material Data**   **CDO**

Crystal chemical data: $[Si_{36}O_{72}]$-**CDO**
orthorhombic, *Pnma*, $a = 18.355$Å, $b = 13.779$Å, $c = 7.3674$Å [1]
(Relationship to unit cell of Framework Type: $a' = b$, $b' = c$, $c' = a$)

Framework density: 19.3 T/1000Å^3

Channels: [010] **8** 3.1 x 4.7* ↔ [001] **8** 2.5 x 4.2*

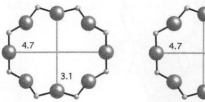 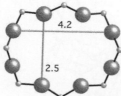 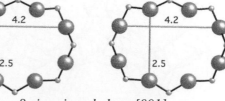

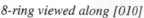

         8-ring viewed along [010]   *8-ring viewed along [001]*

References:
(1) Ikeda, T., Akiyama, Y., Oumi, Y., Kawai, A. and Mizukami, F. *Angew. Chem., Int. Ed.*, **43**, 4892–4896 (2004)
(2) Dorset, D.L. and Kennedy, G.J. *J. Phys. Chem. B*, **108**, 15216–15222 (2004)
(3) Broach, R.W. *private communication*

CFI

Framework Type Data

Imma

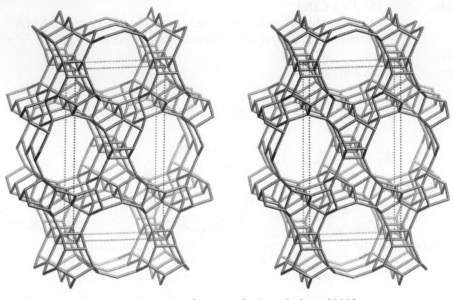

framework viewed along [010]

Idealized cell data: orthorhombic, *Imma*, $a = 14.0$Å, $b = 5.3$Å, $c = 26.0$Å

Coordination sequences and vertex symbols:

T_1 (8,*m*)	4	10	21	36	56	84	114	143	182	231	$4 \cdot 6 \cdot 4 \cdot 6 \cdot 5 \cdot 6$
T_2 (8,*m*)	4	12	23	37	55	83	114	153	195	222	$5 \cdot 6 \cdot 5 \cdot 6 \cdot 5_2 \cdot 6$
T_3 (8,*m*)	4	12	22	37	57	84	114	156	184	222	$5 \cdot 6 \cdot 5 \cdot 6 \cdot 5 \cdot 6_2$
T_4 (4,*mm*2)	4	12	24	36	54	79	118	153	190	234	$5 \cdot 5 \cdot 5 \cdot 5 \cdot 14_{14} \cdot *$
T_5 (4,*mm*2)	4	12	20	34	56	81	116	151	186	220	$5 \cdot 5 \cdot 5 \cdot 5 \cdot 5 \cdot 6_2$

Secondary building units: see *Compendium*

Composite building units:

dzc	*mtt*	*cas*	*ton*
double zigzag chain			

Materials with this framework type:

*CIT-5[1,2]

CIT-5   **Type Material Data**   **CFI**

Crystal chemical data: [Si$_{32}$O$_{64}$]-**CFI**
                               orthorhombic, *Pmn*2$_1$, *a* = 13.674Å, *b* = 5.022Å, *c* = 25.488Å [2]

Framework density: 18.3 T/1000Å^3

Channels: [010] **14** 7.2 x 7.5*

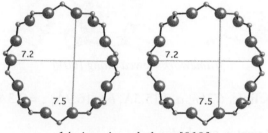

14-ring viewed along [010]

References:
(1) Wagner, P., Yoshikawa, M., Lovallo, M., Tsuji, K., Taspatsis, M. and Davis, M.E. *Chem. Commun.*, 2179–2180 (1997)
(2) Yoshikawa, M., Wagner, P., Lovallo, M., Tsuji, K., Takewaki, T., Chen, C.Y., Beck, L.W., Jones, C., Tsapatsis, M., Zones, S.I. and Davis, M.E. *J. Phys. Chem. B*, **102**, 7139–7147 (1998)

CGF

Framework Type Data

*C*2/*m*

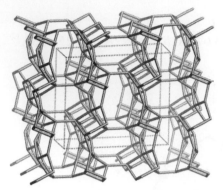 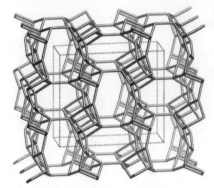

framework viewed along [001]

Idealized cell data: monoclinic, *C*2/*m*, *a* = 15.5Å, *b* = 16.9Å, *c* = 7.3Å, β= 96.1°

Coordination sequences and vertex symbols:

T₁ (8,1)	4	11	22	34	50	76	111	142	165	199	·4·8·6₃·8·6₃·8
T₂ (8,1)	4	10	19	33	55	79	100	129	172	216	4·6·4·6₂·6·6
T₃ (8,1)	4	9	18	34	55	76	97	131	177	217	4·4·4·6·6·6₂
T₄ (8,1)	4	9	18	34	55	75	98	133	177	216	4·6₃·4·6₃·4·8
T₅ (4,2)	4	10	18	32	58	80	96	124	176	228	4·4·6₂·6₂·10·10

Secondary building units: see *Compendium*

Composite building units:

    mei   *bog*

Materials with this framework type:
    *Co-Ga-Phosphate-5[1]
    [Zn-Ga-P-O]-**CGF**[2]

Co-Ga-Phosphate-5　　Type Material Data　　**CGF**

Crystal chemical data:　　$|(C_6H_{14}N_2)_2|\ [Co_4Ga_5P_9O_{36}]$-**CGF**
$C_6H_{12}N_2$ = DABCO
monoclinic, $I2/a$

$a = 15.002\text{Å},\ b = 17.688\text{Å},\ c = 15.751\text{Å},\ \beta = 97.24$ [1]
(Relationship to unit cell of Framework Type: $a' = 2c,\ b' = b,\ c' = a$)

Framework density:　　$17.4\ T/1000\text{Å}^3$

Channels:　　$\{[100]\ \mathbf{10}\ 2.5 \times 9.2^* + \mathbf{8}\ 2.1 \times 6.7^*\} \leftrightarrow [001]\ \mathbf{8}\ 2.4 \times 4.8^*$

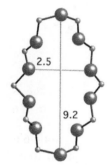 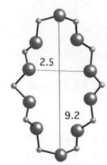 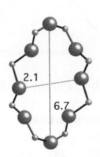 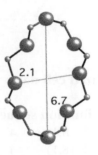

10-ring viewed along [100]　　　　*8-ring viewed along [100]*

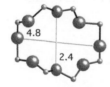 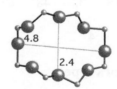

8-ring viewed along [001]

References:
(1) Chippindale, A.M. and Cowley, A.R. *Zeolites*, **18**, 176–181 (1997)
(2) Cowley, A.R., Jones, R.H., Teat, S.J. and Chippindale, A.M. *Microporous Mesoporous Mat.*, **51**, 51–64 (2002)

CGS

Framework Type Data

Pnma

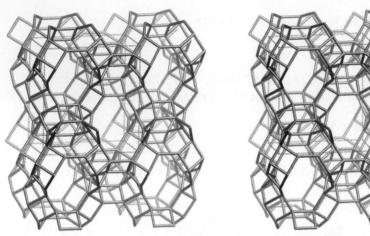

framework viewed along [100]

Idealized cell data: orthorhombic, *Pnma*, $a = 8.4$Å, $b = 14.1$Å, $c = 15.9$Å

Coordination sequences and vertex symbols:

T_1 (8,1)	4	9	16	26	43	67	91	116	148	188	$4 \cdot 4 \cdot 4 \cdot 6 \cdot 6 \cdot 8$
T_2 (8,1)	4	9	18	32	48	66	91	121	150	184	$4 \cdot 4 \cdot 4 \cdot 10_2 \cdot 8 \cdot 8_6$
T_3 (8,1)	4	9	17	28	45	66	91	119	148	186	$4 \cdot 4 \cdot 4 \cdot 8_2 \cdot 6 \cdot 10$
T_4 (8,1)	4	9	18	32	48	67	91	119	151	185	$4 \cdot 4 \cdot 4 \cdot 8 \cdot 8_4 \cdot 10$

Secondary building units: 4

Materials with this framework type:
 *Co-Ga-Phosphate-6[1]
 [Zn-Ga-P-O]-**CGS**[1,2]
 TNU-1, [Ga-Si-O]-**CGS**[3]
 TsG-1, [Ga-Si-O]-**CGS**[4]

Co-Ga-Phosphate-6   Type Material Data   CGS

Crystal chemical data: $|(C_7H_{14}N)_4|\ [Co_4Ga_{12}P_{16}O_{64}]$-**CGS**
$C_7H_{14}N$ = quinuclidinium
monoclinic, $P2_1/c$

$a = 14.365$Å, $b = 16.305$Å, $c = 8.734$Å, $\beta = 90.24$ [1]
(Relationship to unit cell of Framework Type: $a' = b$, $b' = c$, $c' = a$)

Framework density: 15.6 T/1000Å^3

Channels: {[001] **10** 3.5 x 8.1 ↔ [100] **8** 2.5 x 4.6}***

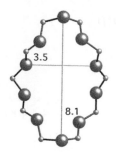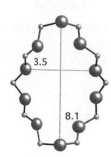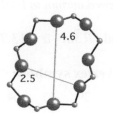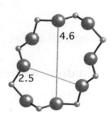

10-ring viewed along [001]   *8-ring viewed along [100]*

References:
(1) Cowley, A.R. and Chippindale, A.M. *Microporous Mesoporous Mat.*, **28**, 163–172 (1999)
(2) Lin, C.-H. and Wang, S.-L. *Chem. Mater.*, **12**, 3617–3623 (2000)
(3) Hong, S.B., Kim, S.H., Kim, Y.G., Kim, Y.C., Barrett, P.A. and Camblor, M.A. *J. Mater. Chem.*, **9**, 2287–2289 (1999)
(4) Lee, Y.J., Kim, S.J., Wu, G. and Parise, J.B. *Chem. Mater.*, **11**, 879–880 (1999)

CHA

Framework Type Data

$R\overline{3}m$

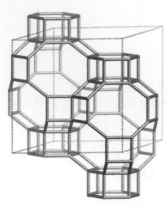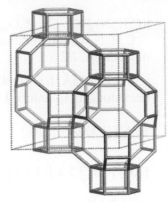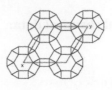

framework viewed normal to [001] (upper right: projection down [001])

Idealized cell data: trigonal, $R\overline{3}m$, $a = 13.7$Å, $c = 14.8$Å

Coordination sequences and vertex symbols:

T₁ (36,1)   4   9   17   29   45   64   85   110   140   173   4·4·4·8·6·8

Secondary building units: 6-6 or 6 or 4-2 or 4

Framework description: AABBCC sequence of 6-rings

Composite building units:

d6r

cha

Materials with this framework type:

*Chabazite[1,2]
[Al-As-O]-**CHA**[3]
[Co-Al-P-O]-**CHA**[4,5]
[Mg-Al-P-O]-**CHA**[5]
[Si-O]-**CHA**[6]
[Zn-Al-P-O]-**CHA**[7]
[Zn-As-O]-**CHA**[3]
|Co|[Be-P-O]-**CHA**[8]
|Li-Na|[Al-Si-O]-**CHA**[9]
AlPO-34[10]

CoAPO-44[11]
CoAPO-47[11]
DAF-5[12]
Dehyd. Na-Chabazite[13]
GaPO-34[14]
K-Chabazite, Iran[15]
LZ-218[16]
Linde D[17,18]
Linde R[19]
MeAPO-47[11,20,21]

MeAPSO-47[11,20,21]
Ni(deta)₂-UT-6[22]
Phi[18,23,24]
SAPO-34[25]
SAPO-47[26]
UiO-21[27]
Willhendersonite[28]
ZK-14[29]
ZYT-6[30]

Chabazite Type Material Data **CHA**

Crystal chemical data: |Ca$_6$ (H$_2$O)$_{40}$| [Al$_{12}$Si$_{24}$O$_{72}$]-**CHA**

rhombohedral, $R\bar{3}m$, a = 9.42Å, α = 94.47° [2]

Framework density: 14.5 T/1000Å^3

Channels: ⊥ [001] **8** 3.8 x 3.8***

(variable due to considerable flexibility of framework)

see Appendix A for 8-ring viewed normal to [001]

References:
(1) Dent, L.S. and Smith, J.V. *Nature*, **181**, 1794–1796 (1958)
(2) Smith, J.V., Rinaldi, F. and Dent Glasser, L.S. *Acta Crystallogr.*, **16**, 45–53 (1963)
(3) Feng, P., Zhang, T. and Bu, X. *J. Am. Chem. Soc.*, **123**, 8608–8609 (2001)
(4) Feng, P., Bu, X. and Stucky, G.D. *Nature*, **388**, 735–741 (1997)
(5) Feng, P., Bu, X., Gier, T.E. and Stucky, G.D. *Microporous Mesoporous Mat.*, **23**, 221–229 (1998)
(6) Díaz-Cabañas, M.-J., Barrett, P.A. and Camblor, M.A. *Chem. Commun.*, 1881–1882 (1998)
(7) Tusar, N.N., Kaucic, V., Geremia, S. and Vlaic, G. *Zeolites*, **15**, 708–713 (1995)
(8) Zhang, H.Y., Weng, L.H., Zhou, Y.M., Chen, Z.X., Sun, J.Y. and Zhao, D.Y. *J. Mater. Chem.*, **12**, 658–662 (2002)
(9) Smith, L.J., Eckert, H. and Cheetham, A.K. *J. Am. Chem. Soc.*, **122**, 1700–1708 (2000)
(10) Harding, M.M. and Kariuki, B.M. *Acta Crystallogr.*, **C50**, 852–854 (1994)
(11) Bennett, J.M. and Marcus, B.K. *Stud. Surf. Sci. Catal.*, **37**, 269–279 (1988)
(12) Sankar, G., Wyles, J.K., Jones, R.H., Thomas, J.M., Catlow, C.R.A., Lewis, D.W., Clegg, W., Coles, S.J. and Teat, S.J. *Chem. Commun.*, 117–118 (1998)
(13) Mortier, W.J., Pluth, J.J. and Smith, J.V. *Mater. Res. Bull.*, **12**, 241–250 (1977)
(14) Schott-Darie, C., Kessler, H., Soulard, M., Gramlich, V. and Benazzi, E. *Stud. Surf. Sci. Catal.*, **84**, 101–108 (1994)
(15) Calligaris, M., Nardin, G. and Randaccio, L. *Zeolites*, **3**, 205–208 (1983)
(16) Breck, D.W. and Skeels, G.W. *U.S. Patent 4,333,859* (1982)
(17) Breck, D.W. and Acara, N.A. *U.S. Patent 2,950,952* (1960)
(18) Lillerud, K.P., Szostak, R. and Long, A. *J. Chem. Soc., Faraday Trans.*, **90**, 1547–1551 (1994)
(19) Milton, R.M. *Brit. Patent 841,812* (1960)
(20) Flanigen, E.M., Lok, B.M., Patton, R.L. and Wilson, S.T. *Pure Appl. Chem.*, **58**, 1351–1358 (1986)
(21) Flanigen, E.M., Lok, B.M., Patton, R.L. and Wilson, S.T. *Proc. 7th Int. Zeolite Conf.*, pp. 103–112 (1986)
(22) Garcia, R., Shannon, I.J., Slawin, A.M.Z., Zhou, W., Cox, P.A. and Wright, P.A. *Microporous Mesoporous Mat.*, **58**, 91–104 (2003)
(23) Grose, R.W. and Flanigen, E.M. *U.S. Patent 4,124,686* (1978)
(24) Lobo, R.F., Annen, M.J. and Davis, M.E. *J. Chem. Soc., Faraday Trans.*, **88**, 2791–2795 (1992)
(25) Lok, B.M., Messina, C.A., Patton, R.L., Gajek, R.T., Cannan, T.R. and Flanigen, E.M. *J. Am. Chem. Soc.*, **106**, 6092–6093 (1984)
(26) Pluth, J.J. and Smith, J.V. *J. Phys. Chem.*, **93**, 6516–6520 (1989)
(27) Kongshaug, K.O., Fjellvåg, H. and Lillerud, K.P. *Microporous Mesoporous Mat.*, **39**, 341–350 (2000)
(28) Tillmanns, E., Fischer, R.X. and Baur, W.H. *N. Jb. Miner. Mh.*, 547–558 (1984)
(29) Kuehl, G.H. *Molecular Sieves*, pp. 85–91 (1968)
(30) Ito, M., Shimoyama, Y., Saito, Y., Tsurita, Y. and Otake, M. *Acta Crystallogr.*, **C41**, 1698–1700 (1985)

-CHI

Pbcn

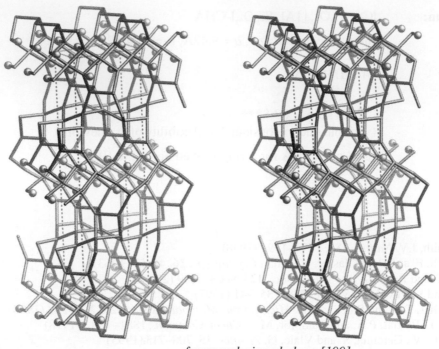

framework viewed along [100]

Idealized cell data: orthorhombic, *Pbcn*, $a = 5.0$Å, $b = 31.2$Å, $c = 9.0$Å

Coordination sequences and vertex symbols:

T_1 (8,1)	4	11	22	40	64	89	120	160	203	248	$4 \cdot 6_3 \cdot 6 \cdot 6 \cdot 6_2 \cdot 6_2$
T_2 (8,1)	4	10	20	36	60	86	115	157	196	238	$4 \cdot 5 \cdot 6 \cdot 9 \cdot 6_3 \cdot 10_2$
T_3 (8,1)	3	8	13	29	53	80	113	147	193	231	$4 \cdot 5 \cdot 9$
T_4 (4,2)	4	6	14	28	56	80	114	152	190	236	$4 \cdot 4 \cdot 5 \cdot 9 \cdot 10 \cdot 10$

Secondary building units: 5-[1,1]

Materials with this framework type:
*Chiavennite[1]

Chiavennite   Type Material Data   -CHI

Crystal chemical data: |Ca$_4$ Mn$_4$ (H$_2$O)$_8$| [Be$_8$Si$_{20}$O$_{52}$ (OH)$_8$]-**CHI**
orthorhombic, *Pnab*, a = 8.729Å, b = 31.326Å, c = 4.903Å [1]
(Relationship to unit cell of Framework Type: a' = -c, b' = b, c' = a)

Framework density: 20.9 T/1000Å^3

Channels: [001] **9** 3.9 x 4.3*

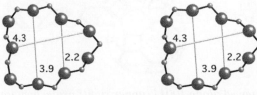

9-ring viewed along [001]

References:
(1)   Tazzoli, V., Domeneghetti, M.C., Mazzi, F. and Cannillo, E. *Eur. J. Mineral.*, **7**, 1339–1344 (1995)

-CLO

Framework Type Data

Pm$\bar{3}$m

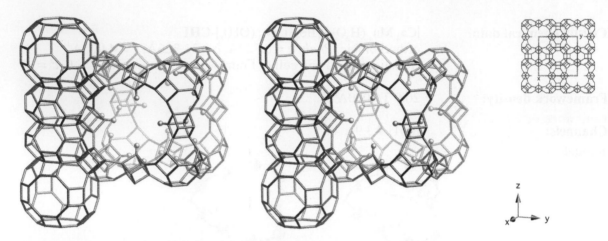

framework viewed along [001] (upper right: projection down [001])

Idealized cell data: cubic, Pm$\bar{3}$m, a = 25.8Å

Coordination sequences and vertex symbols:

T$_1$ (48,1)	4	9	16	23	31	44	59	74	91	109	·4·6·4·6·4·12
T$_2$ (48,1)	4	9	17	27	37	47	56	66	80	99	4·6·4·6·4·8
T$_3$ (48,1)	4	8	13	22	34	44	55	72	94	117	4·6·4·6·4·20$_8$
T$_4$ (24,m)	4	9	16	23	32	45	58	76	98	118	4·6·4·6·4·8
T$_5$ (24,m)	3	5	10	18	29	45	56	65	86	110	4·4·4

Secondary building units: 4-4 or 4

Composite building units:

d4r   clo   lta

  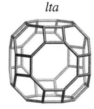

Materials with this framework type:
  *Cloverite[1]
  [Mn-Ga-P-O]--**CLO**[2]
  [Zn-Ga-P-O]--**CLO**[2]

Cloverite   Type Material Data   **-CLO**

Crystal chemical data: |(C$_7$H$_{14}$N)$_{24}$|$_8$ [F$_{24}$Ga$_{96}$P$_{96}$O$_{372}$ (OH)$_{24}$]$_8$-**CLO**
C$_7$H$_{14}$N = quinuclidinium
cubic, $Fm\overline{3}c$, a = 51.712Å [1]
(Relationship to unit cell of Framework Type: $a' = 2a$)

Framework density: 11.1 T/1000Å^3

Channels: <100> **20** 4.0 x 13.2*** | <100> **8** 3.8 x 3.8***

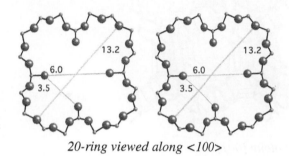
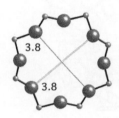

20-ring viewed along <100>

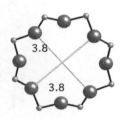

8-ring viewed along <100>

References:
(1)   Estermann, M., McCusker, L.B., Baerlocher, Ch., Merrouche, A. and Kessler, H. *Nature*, **352**, 320–323 (1991)
(2)   Yoshino, M., Matsuda, M. and Miyake, M. *Solid State Ionics*, **151**, 269–274 (2002)

CON

Framework Type Data

C2/m

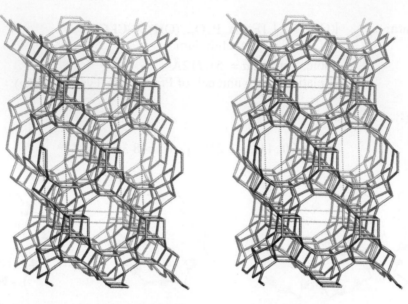

framework viewed along [001]

Idealized cell data: monoclinic, *C2/m*, *a* = 22.7Å, *b* = 13.4Å, *c* = 12.6Å, β= 69.5°

Coordination sequences and vertex symbols:

T_1 (8,1)	4	10	19	32	50	73	101	132	164	199	$4·5·4·10_4·5·6_2$
T_2 (8,1)	4	11	18	31	52	77	98	126	164	205	$4·5·5·6·5·10_2$
T_3 (8,1)	4	10	21	32	47	74	105	134	159	196	$4·6·4·10_4·5·6_2$
T_4 (8,1)	4	10	19	32	51	74	100	130	165	203	$4·4·5·6_2·5·12_7$
T_5 (8,1)	4	10	19	32	51	74	101	130	164	203	$4·4·5·6_2·5·12_4$
T_6 (8,1)	4	11	18	28	49	77	103	126	155	201	$4·5_2·5·6_2·6·6_2$
T_7 (8,1)	4	11	19	32	49	75	105	131	159	196	$4·6_2·5·6·5·10_2$

Secondary building units: 1-5-1

Composite building units:

bea	*bre*	*lau*	*mel*

Materials with this framework type:

*CIT-1[1]   SSZ-26 (related to **CON**)[1,2]   SSZ-33 (related to **CON**)[1,2]

CIT-1   **Type Material Data**   **CON**

Crystal chemical data: $|H_2|$ $[B_2Si_{54}O_{112}]$-**CON**
monoclinic, $C2/m$

$a = 22.624$Å, $b = 13.350$Å, $c = 12.364$Å, $\beta = 68.91°$ [1]

Framework density: 16.1 T/1000Å^3

Channels: [001] **12** 6.4 x 7.0* $\leftrightarrow$ [100] **12** 7.0 x 5.9* $\leftrightarrow$ [010] **10** 5.1 x 4.5*

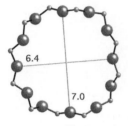 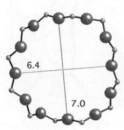 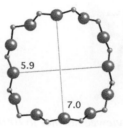 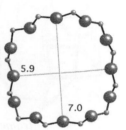

12-ring viewed along [001]   *12-ring viewed along [100]*

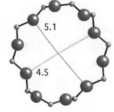 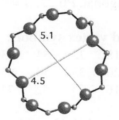

10-ring viewed along [010]

References:
(1) Lobo, R.F. and Davis, M.E. *J. Am. Chem. Soc.*, **117**, 3764–3779 (1995)
(2) Lobo, R.F., Pan, M., Chan, I., Li, H.X., Medrud, R.C., Zones, S.I., Crozier, P.A. and Davis, M.E. *Science*, **262**, 1543–1546 (1993)

CZP

*P*6₁22

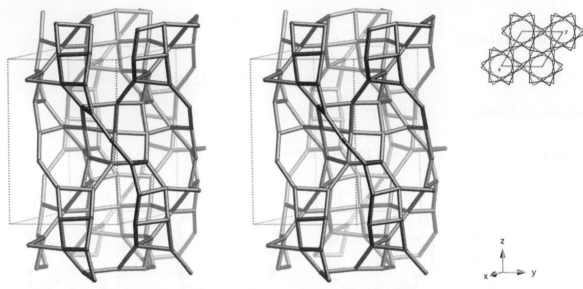

framework viewed normal to [001] (upper right: projection down [001])

Idealized cell data: hexagonal, $P6_1 22$, $a = 9.4$Å, $c = 15.3$Å

Coordination sequences and vertex symbols:

T_1 (12,1)	4	9	18	32	54	83	113	149	191	234	$4 \cdot 4 \cdot 4 \cdot 8_6 \cdot 8 \cdot 8$
T_2 (6,2)	4	10	20	33	56	85	114	144	192	242	$4 \cdot 4 \cdot 8_3 \cdot 8_3 \cdot 8_6 \cdot 8_6$
T_3 (6,2)	4	8	16	33	52	73	112	160	190	214	$4 \cdot 4 \cdot 4 \cdot 4 \cdot 8 \cdot 8$

Secondary building units: 4-[1,1] or 4

Materials with this framework type:
　　*Chiral Zincophosphate[1,2]
　　|(H₃DETA)₂ (H₂O)₁₂|[Mn₆Ga₆P₁₂O₄₈]-**CZP**[3]
　　|Na-|[Co-Zn-P-O]-**CZP**[4]
　　[Zn-B-P-O]-**CZP**[5]

Chiral Zincophosphate   Type Material Data

Crystal chemical data: $|Na_{12} (H_2O)_{12}| [Zn_{12}P_{12}O_{48}]$-**CZP**
hexagonal, $P6_122$, $a = 10.480$Å, $c = 15.089$Å [(2)]

Framework density: 16.7 T/1000Å^3

Channels: [001] **12** 3.8 x 7.2* (highly distorted 12-ring)

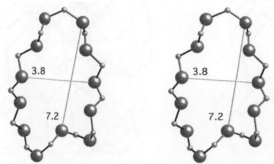

distorted 12-ring viewed along [001]

References:
(1) Rajic, N., Logar, N.Z. and Kaucic, V. *Zeolites*, **15**, 672–678 (1995)
(2) Harrison, W.T.A., Gier, T.E., Stucky, G.D., Broach, R.W. and Bedard, R.A. *Chem. Mater.*, **8**, 145–151 (1996)
(3) Lin, C.H. and Wang, S.L. *Chem. Mater.*, **14**, 96–102 (2002)
(4) Helliwell, M., Helliwell, J.R., Kaucic, V., Logar, N.Z., Barba, L., Busetto, E. and Lausi, A. *Acta Crystallogr.*, **B55**, 327–332 (1999)
(5) Wiebcke, M., Bogershausen and A., Koller, H. *Microporous Mesoporous Mat.*, **78**, 97–102 (2005)

DAC

Framework Type Data

C2/m

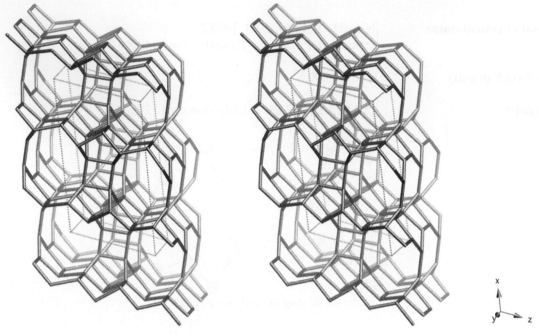

framework viewed along [001]

Idealized cell data: monoclinic, *C2/m*, *a* = 18.6Å, *b* = 7.5Å, *c* = 10.4Å, β = 108.9°

Coordination sequences and vertex symbols:

T_1 (8,1)	4	12	22	39	65	91	121	163	208	250	$5·5·5·5_2·8·10_2$
T_2 (8,1)	4	12	20	37	63	91	118	164	212	245	$5·5·5·5_2·5·8$
T_3 (4,*m*)	4	11	24	41	59	99	130	155	202	262	$4·5_2·5·8·5·8$
T_4 (4,*m*)	4	11	24	39	63	95	132	156	199	266	$4·5_2·5·8·5·8$

Secondary building units: 5-1

Composite building units:

mor

Materials with this framework type:

    *Dachiardite[1,2]

    Svetlozarite (disordered variant, since discredited)[3]

Dachiardite

Type Material Data

DAC

Crystal chemical data: $|(Ca_{0.5},K,Na)_5 (H_2O)_{12}| [Al_5Si_{19}O_{48}]$-**DAC**
monoclinic, $C2/m$

$a = 18.676$Å, $b = 7.518$Å, $c = 10.246$Å, $\beta = 107.87°$ [2]

Framework density: 17.5 T/1000Å³

Channels: [010] **10** 3.4 x 5.3* ↔ [001] **8** 3.7 x 4.8*

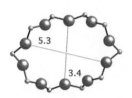 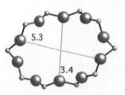

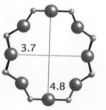 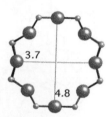

10-ring viewed along [010]   *8-ring viewed along [001]*

References:

(1) Gottardi, G. and Meier, W.M. *Z. Kristallogr.*, **119**, 53–64 (1963)
(2) Vezzalini, G. *Z. Kristallogr.*, **166**, 63–71 (1984)
(3) Gellens, L.R., Price, G.D. and Smith, J.V. *Mineral. Mag.*, **45**, 157–161 (1982)

DDR

Framework Type Data

$R\bar{3}m$

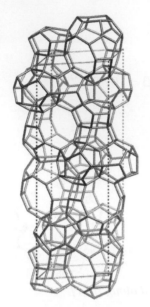 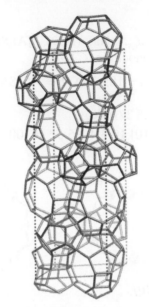 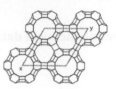

framework viewed normal to [001] (upper right: projection down [001])

Idealized cell data: trigonal, $R\bar{3}m$, a = 13.8Å, c = 40.8Å

Coordination sequences and vertex symbols:

T_1 (36,1)	4	11	23	39	62	91	124	159	203	251	4·5·5·5·5·8
T_2 (18,*m*)	4	12	22	37	59	93	127	158	193	251	5·5·5·5·5·5
T_3 (18,*m*)	4	12	25	40	61	86	119	164	212	253	5·5·5·5·5·6
T_4 (18,*m*)	4	12	24	40	63	87	121	165	208	255	5·5·5·5·5·8
T_5 (18,2)	4	10	21	37	62	94	124	158	196	252	4·4·5·5·6·8
T_6 (6,3*m*)	4	12	24	39	57	93	121	157	210	240	5·5·5·5·5·5
T_7 (6,3*m*)	4	12	24	33	60	97	136	150	192	264	5·5·5·5·5·5

Secondary building units: see *Compendium*

Composite building units:

mtn

Materials with this framework type:
*Deca-dodecasil 3R[1]
[B-Si-O]-**DDR**[2]

Sigma-1[3]
ZSM-58[4,5]

Deca-dodecasil 3R   **Type Material Data**   **DDR**

Crystal chemical data: $|(C_{10}H_{17}N)_6 (N_2)_9| [Si_{120}O_{240}]$-**DDR**
$C_{10}H_{17}N$ = 1-aminoadamantane
trigonal, $R\bar{3}m$, a = 13.860Å, c = 40.891Å [1]

Framework density: 17.6 T/1000Å^3

Channels: $\perp$ [001] **8** 3.6 x 4.4**

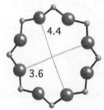 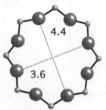

8-ring viewed normal to [001]

References:
(1) Gies, H. *Z. Kristallogr.*, **175**, 93–104 (1986)
(2) Grünewald-Luke, A., Marler, B., Hochgrafe, M. and Gies, H. *J. Mater. Chem.*, **9**, 2529–2536 (1999)
(3) Stewart, A., Johnson, D.W. and Shannon, M.D. *Stud. Surf. Sci. Catal.*, **37**, 57–64 (1988)
(4) Valyocsik, E.W. *U.S. Patent 4,698,217* (1987)
(5) Ernst, S., Chen, C.Y., Lindner, D. and Weitkamp, J. *Zeolites for the Nineties, Recent Progress Reports - Abstracts*, pp. 55–56 (1989)

DFO

Framework Type Data

P6/mmm

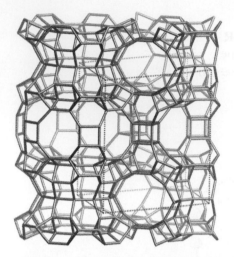 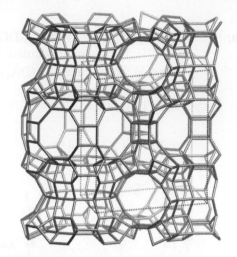 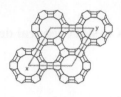

framework viewed normal to [001] (upper right: projection down [001])

Idealized cell data: hexagonal, *P6/mmm*, *a* = 22.0Å, *c* = 21.2Å

Coordination sequences and vertex symbols:

T_1 (24,1)	4	9	16	26	41	60	82	107	135	167	$4 \cdot 6_3 \cdot 4 \cdot 6_3 \cdot 4 \cdot 12$
T_2 (24,1)	4	10	18	27	42	62	84	109	135	167	$4 \cdot 6_2 \cdot 4 \cdot 8 \cdot 6 \cdot 6_2$
T_3 (24,1)	4	10	18	28	45	65	84	106	134	173	$4 \cdot 4 \cdot 6 \cdot 6_3 \cdot 6_3 \cdot 10$
T_4 (24,1)	4	9	17	29	45	63	82	106	136	168	$4 \cdot 6 \cdot 4 \cdot 6_2 \cdot 4 \cdot 8$
T_5 (24,1)	4	9	17	29	44	62	85	112	139	169	$4 \cdot 4 \cdot 4 \cdot 6 \cdot 6 \cdot 6_2$
T_6 (12,*m*)	4	10	18	28	45	66	89	115	141	171	$4 \cdot 6_2 \cdot 4 \cdot 6_2 \cdot 10 \cdot 12$

Secondary building units: see *Compendium*

Composite building units:

d4r   sti   bog   lau

Materials with this framework type:
   *DAF-1[1,2]

DAF-1   Type Material Data   **DFO**

Crystal chemical data: $|(C_{16}H_{38}N_2)_7 (H_2O)_{40}| [Mg_{14}Al_{52}P_{66}O_{264}]$-**DFO**
                          $C_{16}H_{38}N_2$ = decamethonium
                          hexagonal, $P6/mmm$, $a = 22.351$Å, $c = 21.693$Å [1]

Stability: Transforms to AlPO-5 and AlPO-tridymite on heating to 500°C [1]

Framework density: 14.1 T/1000Å^3

Channels: {[001] **12** 7.3 x 7.3 ↔ ⊥ [001] **8** 3.4 x 5.6}*** ↔ {[001] **12** 6.2 x 6.2

              ↔ ⊥ [001] **10** 5.4 x 6.4}***

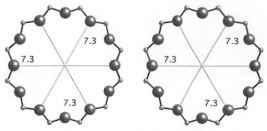

12-ring viewed along [001]

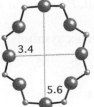

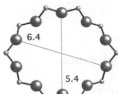

8-ring viewed normal to [001]

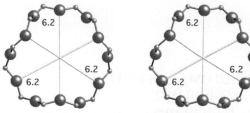

2nd 12-ring viewed along [001]

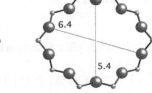

10-ring viewed normal to [001]

References:
(1)   Wright, P.A., Jones, R.H., Natarajan, S., Bell, R.G., Chen, J.S., Hursthouse, M.B. and Thomas, J.M. *Chem. Commun.*, 633–635 (1993)
(2)   Muncaster, G., Sankar, G., Catlow, C.R.A., Thomas, J.M., Bell, R.G., Wright, P.A., Coles, S., Teat, S.J., Clegg, W. and Reeve, W. *Chem. Mater.*, **11**, 158–163 (1999)

DFT

Framework Type Data

$P4_2/mmc$

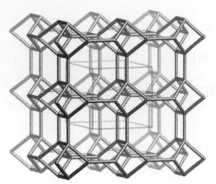 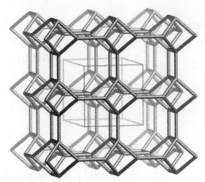

framework viewed along [001]

Idealized cell data: tetragonal, $P4_2/mmc$, $a = 7.1$Å, $c = 9.0$Å

Coordination sequences and vertex symbols:

T_1 (8,m)	4	10	21	36	55	79	106	138	175	215

$4 \cdot 4 \cdot 6_2 \cdot 8_3 \cdot 6_2 \cdot 8_3$

Secondary building units: 4

Composite building units:
> *nsc*
> *narsarsukite*
> *chain*

Materials with this framework type:

*DAF-2[1]
ACP-3 ([Co-Al-P-O]-**DFT**)[2]
UCSB-3GaGe[3]
UCSB-3ZnAs[2]

UiO-20 ([Mg-P-O]-**DFT**)[4]
[Fe-Zn-P-O]-**DFT**[5]
[Zn-Co-P-O]-**DFT**[6]

Crystal chemical data:

$|(C_2H_{10}N_2)_2|$ $[Co_4P_4O_{16}]$-**DFT**
$C_2H_{10}N_2$ = ethylenediammonium
monoclinic, $I2/b$

$a = 14.719$Å, $b = 14.734$Å, $c = 17.891$Å, $\gamma = 90.02°$ [1]
(Relationship to unit cell of Framework Type:
$a' = 2a$, $b' = 2a$, $c' = 2c$)

Framework density: 16.5 T/1000Å^3

Channels: [001] **8** 4.1 x 4.1* ↔ [100] **8** 1.8 x 4.7* ↔ [010] **8** 1.8 x 4.7*

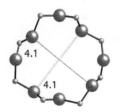 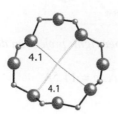 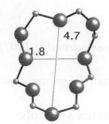 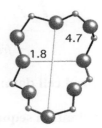

8-ring viewed along [001]   *8-ring viewed along [100]*

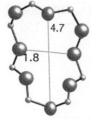 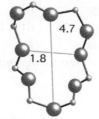

8-ring viewed along[010]

References:

(1) Chen, J., Jones, R.H., Natarajan, S., Hursthouse, M.B. and Thomas, J.M. *Angew. Chem., Int. Ed.*, **33**, 639–640 (1994)
(2) Bu, X., Feng, P., Gier, T.E. and Stucky, G.D. *J. Solid State Chem.*, **136**, 210–215 (1998)
(3) Bu, X., Feng, P., Gier, T.E., Zhao, D. and Stucky, G.D. *J. Am. Chem. Soc.*, **120**, 13389–13397 (1998)
(4) Kongshaug, K.O., Fjellvag, H. and Lillerud, K.P. *Chem. Mater.*, **12**, 1095–1099 (2000)
(5) Zhao, Y.N., Jin, J. and Kwon, Y.U. *Bull. Korean Chem. Soc.*, **26**, 1277–1280 (2005)
(6) Ke, Y., He, G., Li, J., Zhang, Y., Lu, S. and Lei, Z. *Cryst. Res. Technol.*, **37**, 803–811 (2002)

DOH

Framework Type Data

P6/mmm

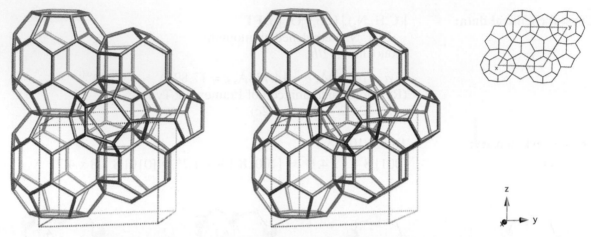

framework viewed normal to [001] (upper right: projection down [001])

Idealized cell data: hexagonal, *P6/mmm*, *a* = 14.2Å, *c* = 11.5Å

Coordination sequences and vertex symbols:

T₁ (12,*m*)	4	12	23	41	64	92	128	167	207	259	5·5·5·5·5·6
T₂ (12,*m*)	4	11	24	41	63	91	128	171	214	259	4·5·5·6·5·6
T₃ (6,*mm*2)	4	12	25	42	68	90	122	167	210	268	5·5·5·5·5·6
T₄ (4,3*m*)	4	12	24	36	61	101	133	156	204	256	5·5·5·5·5·5

Secondary building units: see *Compendium*

Composite building units:
  mtn

Materials with this framework type:
  *Dodecasil 1H[1]
  [B-Si-O]-**DOH**[2]

Dodecasil 1H   Type Material Data   DOH

Crystal chemical data: $|C_5H_{11}N\,(N_2)_5|\,[Si_{34}O_{68}]$-**DOH**
$C_5H_{11}N$ = piperidine
hexagonal, $P6/mmm$, $a = 13.783$Å, $c = 11.190$Å [1]

Framework density: 18.5 T/1000Å^3

Channels: apertures formed by 6-rings only

References:
(1) Gerke, H. and Gies, H. *Z. Kristallogr.*, **166**, 11–22 (1984)
(2) Grünewald-Luke, A., Marler, B., Hochgrafe, M. and Gies, H. *J. Mater. Chem.*, **9**, 2529–2536 (1999)

DON

Cmcm

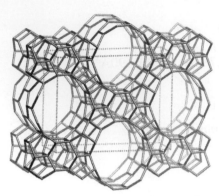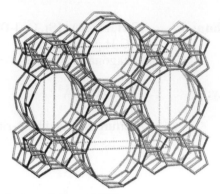

framework viewed along [001]

Idealized cell data: orthorhombic, *Cmcm*, $a = 18.9$Å, $b = 23.4$Å, $c = 8.5$Å

Coordination sequences and vertex symbols:

T_1 (16,1)	4	10	20	34	54	77	107	140	175	218	$4·6·4·6_2·5·6$
T_2 (16,1)	4	10	20	35	54	77	106	138	177	221	$4·6·4·6_2·5·6$
T_3 (16,1)	4	12	24	38	55	76	105	143	184	223	$5·6_2·6·6_2·6·6_2$
T_4 (8,m)	4	12	22	33	53	80	109	143	179	217	$5·6·5·6·6·6_2$
T_5 (8,m)	4	12	23	37	52	74	107	143	183	223	$5·6_2·5·6_2·6_2·6_2$

Secondary building units: 5-3

Composite building units:

dcc	*afi*	*mel*
double crankshaft chain 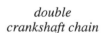		

Materials with this framework type:
  *UTD-1F[1]
  UTD-1[2]

UTD-1F | **Type Material Data** | **DON**

Crystal chemical data: $|((Cp^*)_2Co)_2 F_{1.5}(OH)_{0.5}| [Si_{64}O_{128}]$-**DON**
Cp^* = pentamethylcyclopentadiene
monoclinic, Pc

$a = 14.970$Å, $b = 8.476$Å, $c = 30.028$Å, $\beta = 102.65°$ [1]
(Relationship to unit cell of Framework Type:

$a' = a/(2\sin\beta')$, $b' = c$, $c' = \sqrt{(a^2+c^2)}$
or, as vectors, $\mathbf{a}' = (\mathbf{a} - \mathbf{b})/2$, $\mathbf{b}' = \mathbf{c}$, $\mathbf{c}' = \mathbf{a} + \mathbf{b}$)

Framework density: 17.2 T/1000Å^3

Channels: [010] **14** 8.1 x 8.2*

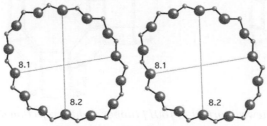

14-ring viewed along [010]

References:
(1) Wessels, T., Baerlocher, C., McCusker, L.B. and Creyghton, E.J. *J. Am. Chem. Soc.*, **121**, 6242–6247 (1999)
(2) Lobo, R.F., Tsapatsis, M., Freyhardt, C.C., Khodabandeh, S., Wagner, P., Chen, C.Y., Balkus, K.J., Zones, S.I. and Davis, M.E. *J. Am. Chem. Soc.*, **119**, 8474–8484 (1997)

EAB

Framework Type Data

$P6_3/mmc$

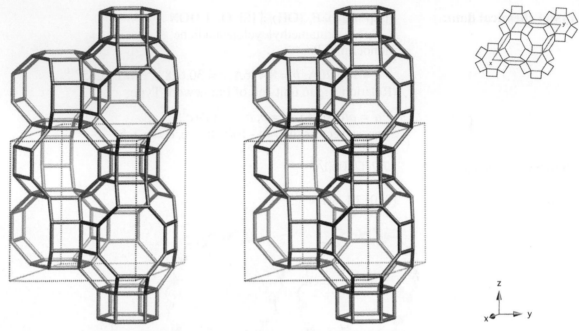

framework viewed normal to [001] (upper right: projection down [001])

Idealized cell data: hexagonal, $P6_3/mmc$, a = 13.2Å, c = 15.0Å

Coordination sequences and vertex symbols:

T_1 (24,1)	4	9	17	30	49	71	92	115	147	190	4·4·4·6·6·8
T_2 (12,2)	4	10	20	32	46	66	94	128	162	192	4·4·6·6·8·8

Secondary building units: 6 or 4

Framework description: ABBACC sequence of 6-rings

Composite building units:

     d6r   *gme*

Materials with this framework type:
    *TMA-E (Aiello and Barrer)[1,2]
    Bellbergite[3]

TMA-E (Aiello and Barrer)   Type Material Data   **EAB**

Crystal chemical data: |$(C_4H_{12}N)_2Na_7 (H_2O)_{26}$| $[Al_9Si_{27}O_{72}]$-**EAB**
$C_4H_{12}N$ = tetramethylammonium
hexagonal, $P6_3/mmc$, $a = 13.28$Å, $c = 15.21$Å [2]

Framework density: 15.5 T/1000Å^3

Channels: $\perp$ [001] **8** 3.7 x 5.1**

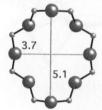 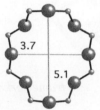

8-ring viewed normal to [001]

References:
(1) Aiello, R. and Barrer, R.M. *J. Chem. Soc. (A)*, 1470–1475 (1970)
(2) Meier, W.M. and Groner, M. *J. Solid State Chem.*, **37**, 204–218 (1981)
(3) Rüdinger, B., Tillmanns, E. and Hentschel, G. *Miner. Petrol.*, **48**, 147–152 (1993)

EDI

Framework Type Data

$P\overline{4}m2$

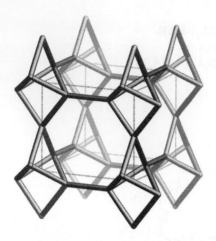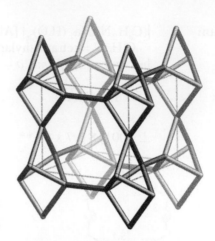

framework viewed normal to [001]

Idealized cell data: tetragonal, $P\overline{4}m2$, a = 6.9Å, c = 6.4Å

Coordination sequences and vertex symbols:

T_1 (4,m)	4	9	19	35	52	72	100	131	163	201	$4 \cdot 8_3 \cdot 4 \cdot 8_3 \cdot 4_2 \cdot 8_4$
T_2 (1, $\overline{4}m2$)	4	8	18	32	52	74	100	128	162	204	$4_2 \cdot 4_2 \cdot 8_4 \cdot 8_4 \cdot 8_4 \cdot 8_4$

Secondary building units: 4=1

Composite building units:

nat

Materials with this framework type:

*Edingtonite[1-3]

[Co-Al-P-O]-**EDI**[4]

[Co-Ga-P-O]-**EDI**[4]

[Zn-As-O]-**EDI**[5]

|(C$_3$H$_{12}$N$_2$)$_{2.5}$|[Zn$_5$P$_5$O$_{20}$]-**EDI**[6]

|Li-|[Al-Si-O]-**EDI**[7]

|Rb$_7$ Na (H$_2$O)$_3$|[Ga$_8$Si$_{12}$O$_{40}$]-**EDI**[8]

K-F[9,10]

Linde F[11]

Orthorhombic edingtonite[12]

Synthetic edingtonite[13]

Tetragonal edingtonite[14]

Zeolite N[15]

| Edingtonite | Type Material Data | **EDI** |

Crystal chemical data: |Ba$_2$ (H$_2$O)$_8$| [Al$_4$Si$_6$O$_{20}$]-**EDI**
orthorhombic, $P2_12_12$, a = 9.550Å, b = 9.665Å, c = 6.523Å [2]
(Relationship to unit cell of Framework Type:

$a' = a\sqrt{2}, b' = b\sqrt{2}, c' = c$
or, as vectors, **a' = a + b, b' = b - a, c' = c**)

Framework density: 16.6 T/1000Å^3

Channels: <110> **8** 2.8 x 3.8** ↔ [001] **8** 2.0 x 3.1*
(variable due to considerable flexibility of framework)

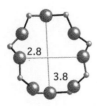 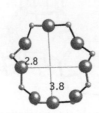 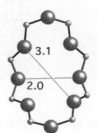 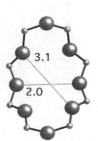

References:
(1) Taylor, W.H. and Jackson, R. *Z. Kristallogr.*, **86**, 53–64 (1933)
(2) Galli, E. *Acta Crystallogr.*, **B32**, 1623–1627 (1976)
(3) Kvick, A. and Smith J.V. *J. Chem. Phys.*, **79**, 2356–2362 (1983)
(4) Bu, X., Gier, T.E., Feng, P. and Stucky, G.D. *Chem. Mater.*, **10**, 2546–2551 (1998)
(5) Feng, P., Zhang, T. and Bu, X. *J. Am. Chem. Soc.*, **123**, 8608–8609 (2001)
(6) Harrison, W.T.A. *Acta Crystallogr.*, **E57**, m248-m250 (2001)
(7) Matsumoto, T., Miyazaki, I. and Goto, Y. *J. Eur. Ceramic Soc.*, **26**, 455–458 (2006)
(8) Lee, Y.J., Kim, S.J. and Parise, J.B. *Microporous Mesoporous Mat.*, **34**, 255–271 (2000)
(9) Barrer, R.M. and Baynham, J.W. *J. Chem. Soc.*, 2882–2891 (1956)
(10) Baerlocher, Ch. and Barrer, R.M. *Z. Kristallogr.*, **140**, 10–26 (1974)
(11) Sherman, J.D. *ACS Sym. Ser.*, **40**, 30–42 (1977)
(12) Gatta, G.D. and Ballaran, T.B. *Mineral. Mag.*, **68**, 167–175 (2004)
(13) Ghobarkar, H. and Schaef, O. *Cryst. Res. Technol.*, **32**, 653–657 (1997)
(14) Mazzi, F., Galli, E. and Gottardi, G. *N. Jb. Miner. Mh.*, 373–382 (1984)
(15) Christensen, A.N. and Fjellvåg, H. *Acta Chemica Scand.*, **51**, 969–973 (1997)

EMT

Framework Type Data

$P6_3/mmc$

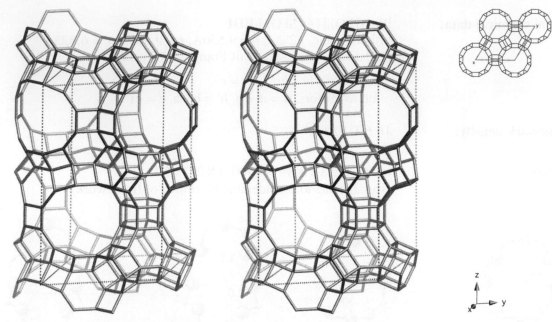

framework viewed normal to [001] (upper right: projection down [001])

Idealized cell data: hexagonal, $P6_3/mmc$, $a = 17.2$Å, $c = 28.1$Å

Coordination sequences and vertex symbols:

T_1 (24,1)	4	9	16	25	37	53	73	96	121	148	178	212	4·4·4·6·6·12
T_2 (24,1)	4	9	16	25	37	53	73	96	121	148	179	214	4·4·4·6·6·12
T_3 (24,1)	4	9	16	25	37	53	73	97	124	152	180	210	4·4·4·6·6·12
T_4 (24,1)	4	9	16	25	37	53	73	96	120	145	174	210	4·4·4·6·6·12

Secondary building units: 6-6 or 6-2 or 6 or 4-2 or 1-4-1 or 4

Composite building units:

d6r

sod

Materials with this framework type:

*EMC-2[1,2]

CSZ-1 (**EMT-FAU** structural intermediate)[3]

ECR-30 (**EMT-FAU** structural intermediate)[4]

ZSM-20 (**EMT-FAU** structural intermediate)[5]

ZSM-3 (**EMT-FAU** structural intermediate)[6]

EMC-2   **Type Material Data**   **EMT**

Crystal chemical data: $|Na_{21} (C_{12}H_{24}O_6)_4| [Al_{21}Si_{75}O_{192}]$-**EMT**
                             $C_{12}H_{24}O_6$ = 18-crown-6
                             hexagonal, $P6_3/mmc$, a = 17.374Å, c = 28.365Å $^{(2)}$

Framework density: 12.9 T/1000Å^3

Channels: [001] **12** 7.3 x 7.3* ↔ ⊥ [001] **12** 6.5 x 7.5**

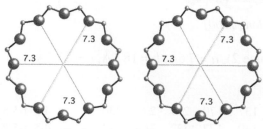 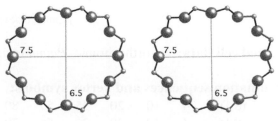

       12-ring viewed along [001]   *12-ring viewed normal to [001]*

References:
(1)   Delprato, F., Delmotte, L., Guth, J.L. and Huve, L. *Zeolites*, **10**, 546–552 (1990)
(2)   Baerlocher, Ch., McCusker, L.B. and Chiappetta, R. *Microporous Materials*, **2**, 269–280 (1994)
(3)   Barrett, M.G. and Vaughan, D.E.W. *UK Patent GB 2,076,793 A* (1981)
(4)   Vaughan, D.E.W. *E. Patent 0,351,461* (1989)
(5)   Newsam, J.M., Treacy, M.M.J., Vaughan, D.E.W., Strohmaier, K.G. and Mortier, W.J. *Chem. Commun.*, 493–495 (1989)
(6)   Kokotailo, G.T. and Ciric, J. *Adv. Chem. Ser.*, **101**, 109–121 (1971)

EON

Framework Type Data

Pmmn

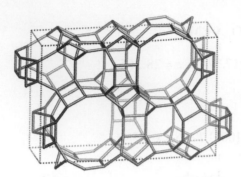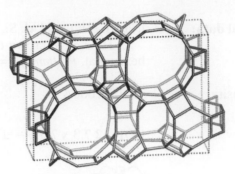

framework viewed along [100]

Idealized cell data: orthorhombic, *Pmmn* (origin choice 2), $a = 7.6$Å, $b = 18.1$Å, $c = 25.9$Å

Coordination sequences and vertex symbols:

T_1 (8,1)	4	10	20	35	55	80	108	137	170	214	272	330	4·5·4·5·8·12
T_2 (8,1)	4	10	20	36	56	79	106	140	177	222	267	317	4·5·4·5·8·12
T_3 (8,1)	4	10	21	37	55	80	111	137	183	232	274	316	4·5·4·5·8·12
T_4 (8,1)	4	12	21	36	59	85	108	150	192	231	277	331	5·5·5·5$_2$·8·12
T_5 (8,1)	4	12	20	36	62	84	112	151	193	229	272	340	5·5·5·5$_2$·5·8
T_6 (4,*m*..)	4	10	21	36	53	77	109	142	175	217	266	315	4·8$_2$·4·8$_2$·5·6
T_7 (4,*m*..)	4	10	21	36	54	76	108	142	181	226	258	309	4·8$_2$·4·8$_2$·5·6
T_8 (4,*m*..)	4	10	21	37	55	76	103	149	191	218	266	326	4·8$_2$·4·8$_2$·5·6
T_9 (4,*m*..)	4	11	24	38	52	82	124	146	182	238	283	325	4·5$_2$·5·8$_2$·5·8$_2$
T_{10} (4,*m*..)	4	11	24	39	57	87	118	149	187	235	282	327	4·5$_2$·5·8·5·8

Secondary building units: 5-1

Composite building units:

dsc	*mor*	*gme*
double sawtooth chain		

Materials with this framework type:

*ECR-1[1-3]

TNU-7[4]

ECR-1   **Type Material Data**   **EON**

Crystal chemical data: |Na$_x$ (H$_2$O)$_y$| [Al$_x$Si$_{60-x}$O$_{120}$]-**EON**, x = 3.6 ... 11.4
orthorhombic, *Pmmn*, *a* = 7.579Å, *b* = 18.089Å, *c* = 25.853Å [4]

Framework density: 16.9 T/1000Å^3

Channels: {[100] **12** 6.7 x 6.8* ↔ [010] **8** {[001] 3.4 x 4.9 ↔
[100] **8** 2.9 x 2.9}*}**
(Two independent, 2-dimensional channel systems, each consisting of
a 12-ring in the [100] direction and a series of 8-rings with effective
diffusion in the [010] direction)

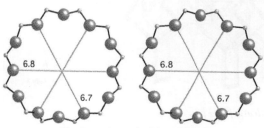

12-ring viewed along [100]

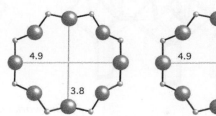

8-ring viewed along [001]

References:
(1) Leonowicz, M.E. and Vaughan, D.E.W. *Nature*, **329**, 819–821 (1987)
(2) Chen, C.S.H., Schlenker, J.L. and Wentzek, S.E. *Zeolites*, **17**, 393–400 (1996)
(3) Gualtieri, A.F., Ferrari, S., Galli, E., Di Renzo, F. and van Beek, W. *Chem. Mater.*, **18**, 76–84 (2006)
(4) Warrender, S.J., Wright, P.A. Zhou, W.Z., Lightfoot, P., Camblor, M.A., Shin, C.H., Kim, D.J. and Hong, S.B. *Chem. Mater.*, **17**, 1272–1274 (2005)

EPI

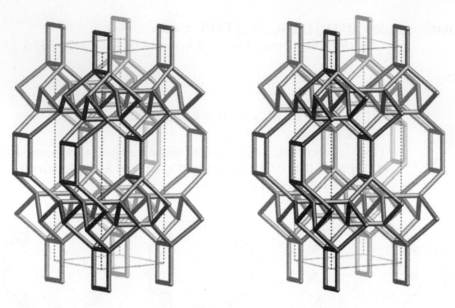

framework viewed along [001]

Idealized cell data: monoclinic, *C2/m*, *a* = 9.1Å, *b* = 17.5Å, *c* = 10.4Å, β = 124.9°

Coordination sequences and vertex symbols:

T_1 (8,1)	4	11	24	42	63	93	127	160	206	262		$4 \cdot 5_2 \cdot 5 \cdot 8 \cdot 5 \cdot 8$
T_2 (8,1)	4	12	22	37	64	94	119	161	204	252		$5 \cdot 5 \cdot 5 \cdot 5_2 \cdot 8 \cdot 10_2$
T_3 (8,1)	4	12	20	39	66	90	118	164	214	245		$5 \cdot 5 \cdot 5 \cdot 5_2 \cdot 5 \cdot 8$

Secondary building units: 5-1

Composite building units:

     mor

Materials with this framework type:
    *Epistilbite[1-4]
    Synthetic epistilbite[5]

Epistilbite   Type Material Data   **EPI**

Crystal chemical data: |Ca$_3$ (H$_2$O)$_{16}$| [Al$_6$Si$_{18}$O$_{48}$]-**EPI**
monoclinic, $C2/m$

$a = 9.08$Å, $b = 17.74$Å, $c = 10.25$Å, $\beta = 124.54°$ [2]

Framework density: 17.6 T/1000Å^3

Channels: {[001] **8** 3.7 x 4.5 $\leftrightarrow$ [100] **8** 3.6 x 3.6}**
(The 8-ring along [100] is tilted and has its normal along [101].

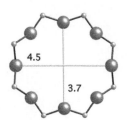 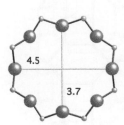 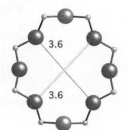 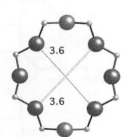

8-ring viewed along [001]   *8-ring viewed along [101]*

References:
(1)   Kerr, I.S. *Nature*, **202**, 589 (1964)
(2)   Perrotta, A.J. *Mineral. Mag.*, **36**, 480–490 (1967)
(3)   Alberti, A., Galli, E. and Vezzalini, G. *Z. Kristallogr.*, **173**, 257–265 (1985)
(4)   Yang, P. and Armbruster, T. *Eur. J. Mineral.*, **8**, 263–271 (1996)
(5)   Ghobarkar, H. *Cryst. Res. Technol.*, 151–1573 (1984)

ERI

Framework Type Data

P6₃/mmc

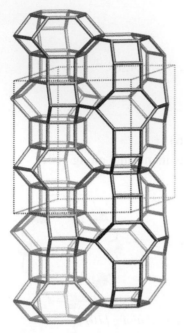
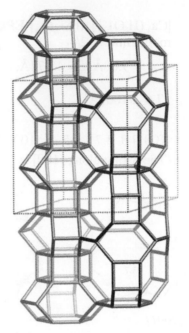
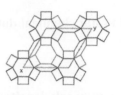

framework viewed normal to [001] (upper right: projection down [001])

Idealized cell data: hexagonal, *P6₃/mmc*, a = 13.1Å, c = 15.2Å

Coordination sequences and vertex symbols:

T₁ (24,1)	4	9	17	30	50	75	98	118	144	185		4·4·4·6·6·8
T₂ (12,*m*)	4	10	20	32	46	64	90	126	164	196		4·8·4·8·6·6

Secondary building units: 6 or 4

Framework description: AABAAC sequence of 6-rings

Composite building units:
    d6r   *can*

Materials with this framework type:
  *Erionite[1-4]
  AlPO-17 plus compositional variants[5-7]

Linde T (**ERI-OFF** structural intermediate)[8]
LZ-220[9]

Erionite   Type Material Data   **ERI**

Crystal chemical data: $|(Ca,Na_2)_{3.5} K_2 (H_2O)_{27}| [Al_9Si_{27}O_{72}]$-**ERI**

hexagonal, $P6_3/mmc$, $a = 13.27$Å, $c = 15.05$Å [3]

Framework density: $15.7 \ T/1000$Å^3

Channels: $\perp$ [001] **8** 3.6 x 5.1***

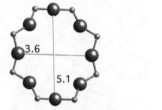 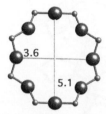

8-ring viewed normal to [001]

References:

(1)   Staples, L.W. and Gard, J.A. *Mineral. Mag.*, **32**, 261–281 (1959)
(2)   Kawahara, A. and Curien, H. *Bull. Soc. fr. Minéral. Cristallogr.*, **92**, 250–256 (1969)
(3)   Gard, J.A. and Tait, J.M. *Proc. 3rd Int. Conf. Molecular Sieves*, pp. 94–99 (1973)
(4)   Schlenker, J.L., Pluth, J.J. and Smith, J.V. *Acta Crystallogr.* **B33**, 3265–3268 (1977)
(5)   Pluth, J.J., Smith, J.V. and Bennett, J.M. *Acta Crystallogr.*, **C42**, 283–286 (1986)
(6)   Flanigen, E.M., Lok, B.M., Patton, R.L. and Wilson, S.T. *Pure Appl. Chem.*, **58**, 1351–1358 (1986)
(7)   Flanigen, E.M., Lok, B.M., Patton, R.L. and Wilson, S.T. *Proc. 7th Int. Zeolite Conf.*, pp. 103–112 (1986)
(8)   Breck, D.W. *Zeolite Molecular Sieves*, p. 173 (1974)
(9)   Breck, D.W. and Skeels, G.W. *U.S. Patent 4,503,023* (1985)

ESV

Framework Type Data

Pnma

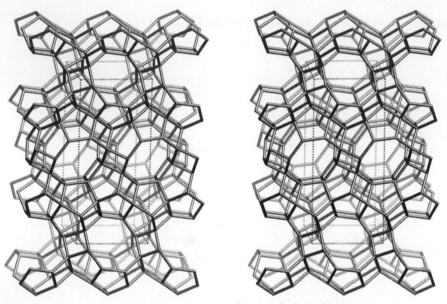

framework viewed along [010]

Idealized cell data: orthorhombic, *Pnma*, a = 9.7Å, b = 12.2Å, c = 22.8Å

Coordination sequences and vertex symbols:

T_1 (8,1)	4	10	20	35	56	82	111	143	180	228	4·5·4·5·6·8
T_2 (8,1)	4	10	20	36	58	82	109	144	186	230	4·4·5·6·6·8
T_3 (8,1)	4	10	21	36	57	82	111	145	183	231	4·4·5·8·6·6
T_4 (8,1)	4	10	21	36	56	82	113	145	180	224	4·4·5·6·6·8
T_5 (8,1)	4	11	22	36	56	78	110	148	184	225	4·6·5·6·5·6
T_6 (8,1)	4	11	20	37	54	82	112	142	182	226	4·5·5·5·6·6

Secondary building units: 5-1

Materials with this framework type:

*ERS-7[1-3]

ERS-7   **Type Material Data**   **ESV**

Crystal chemical data: $|H_{5.06}Na_{0.07}|$ $[Al_{5.13}Si_{42.87}O_{96}]$-**ESV**
orthorhombic, *Pnma*, $a = 9.800$ Å, $b = 12.412$ Å, $c = 22.861$ Å [2]

Framework density: 17.3 T/1000Å^3

Channels: [010] **8** 3.5 x 4.7*

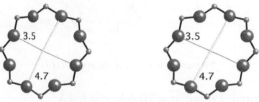

8-ring viewed along [010]

References:
(1) Campbell, B.J., Bellussi, G., Carluccio, L., Perego, G., Cheetham, A.K., Cox, D.E. and Millini, R. *Chem. Commun.*, 1725–1726 (1998)
(2) Millini, R., Perego, G., Carluccio, L., Bellussi, G., Cox, D.E., Campbell, B.J. and Cheetham, A.K. *Proc. 12th Int. Zeolite Conf.*, **I**, pp. 541–548 (1999)
(3) Campbell, B.J., Cheetham, A.K., Vogt, T., Carluccio, L., Parker, W.O., Flego, C. and Millini, R. *J. Phys. Chem. B*, **105**, 1947–1955 (2001)

ETR

Framework Type Data

$P6_3mc$

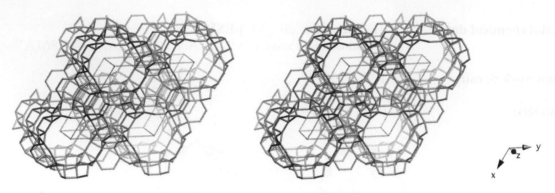

framework viewed along [001]

Idealized cell data: hexagonal, $P6_3mc$, $a = 20.6$Å, $c = 8.4$Å

Coordination sequences and vertex symbols:

T_1 (12,1)	4	9	18	31	45	63	86	111	140	172	210	253	$4 \cdot 4 \cdot 4 \cdot 8 \cdot 8_4 \cdot 10_3$
T_2 (12,1)	4	9	17	28	41	57	80	113	151	189	226	260	$4 \cdot 4 \cdot 4 \cdot 6 \cdot 8 \cdot 8_6$
T_3 (12,1)	4	9	17	28	43	62	83	109	138	170	214	255	$4 \cdot 4 \cdot 4 \cdot 8_2 \cdot 6 \cdot 10_4$
T_4 (12,1)	4	9	16	25	39	58	80	107	143	185	220	254	$4 \cdot 4 \cdot 4 \cdot 6 \cdot 6 \cdot 8$

Secondary building units: 4

Materials with this framework type:
    *ECR-34[1]

ECR-34 **Type Material Data** **ETR**

Crystal chemical data: |H$_{1.2}$K$_{6.3}$Na$_{4.4}$| [Ga$_{11.6}$Al$_{0.3}$Si$_{36.1}$O$_{96}$]-**ETR**
hexagonal, *P6$_3$mc*, *a* = 21.030 Å, *c* = 8.530 Å [1]

Framework density: 14.7 T/1000Å^3

Channels: [001] **18** 10.1* ↔ ⊥ [001] **8** 2.5 x6.0**

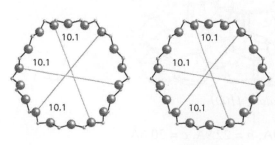 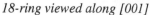 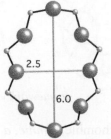 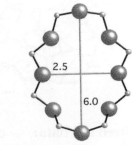

18-ring viewed along [001] *8-ring viewed normal to [001]*

References:
(1) Strohmaier, K. G. and Vaughan, D.E.W. *J. Am. Chem. Soc.*, **125**, 16035–16039 (2003)

EUO

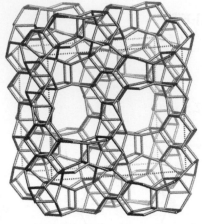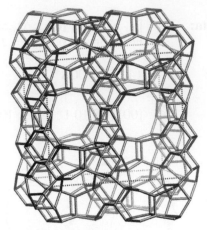

framework viewed along [100]

Idealized cell data: orthorhombic, *Cmme*, a = 13.9Å, b = 22.9Å, c = 20.6Å

Coordination sequences and vertex symbols:

T_1 (16,1)	4	11	21	36	59	92	129	167	197	246	$4·6·5·5·5_2·10$
T_2 (16,1)	4	12	25	41	64	88	122	160	202	256	$5·6·5·6·5·6$
T_3 (16,1)	4	12	23	38	60	89	124	159	194	248	$5·5_2·5·6·5·6$
T_4 (16,1)	4	12	22	39	59	91	124	160	206	257	$5·5·5·6_2·5·10$
T_5 (8,m)	4	12	20	31	61	88	120	159	197	248	$5·5·5·5·5·6_2$
T_6 (8,m)	4	12	20	34	57	92	131	164	202	236	$5·5_2·5·5_2·12_2·*$
T_7 (8,m)	4	12	24	39	60	91	126	161	195	243	$5·6·5·6·5·6_2$
T_8 (8,m)	4	12	24	36	56	90	127	157	197	236	$5·5·5·5·12_4·*$
T_9 (8,m)	4	11	23	45	67	88	115	162	218	261	$4·6·5·5·5·5$
T_{10} (8,m)	4	11	24	39	68	92	118	156	213	268	$4·10·5·5·5·5$

Secondary building units: 1-5-1

Composite building units:

cas   *non*

Materials with this framework type:
*EU-1[1,2]   ZSM-50[6]
[B-Si-O]-**EUO**[3,4]   o-FDBDM-ZSM-50[7]
TPZ-3[5]

EU-1   **Type Material Data**   **EUO**

Crystal chemical data: $|Na_n (H_2O)_{26}| [Al_nSi_{112-n}O_{224}]$-**EUO** , n < 19, typically n ~ 3.6
orthorhombic, *Cmme*, a = 13.695Å, b = 22.326Å, c = 20.178Å [(2)]

Framework density: 18.2 T/1000Å^3

Channels: [100] **10** 4.1 x 5.4* with large side pockets

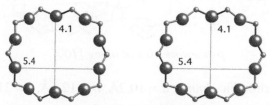

10-ring viewed along [100]

References:
(1) Casci, J.L., Lowe, B.M. and Whittam, T.V. *U.S. Patent 4,537,754* (1985)
(2) Briscoe, N.A., Johnson, D.W., Shannon, M.D., Kokotailo, G.T. and McCusker, L.B. *Zeolites*, **8**, 74–76 (1988)
(3) Grünewald-Luke, A., Marler, B., Hochgrafe, M. and Gies, H. *J. Mater. Chem.*, **9**, 2529–2536 (1999)
(4) Millini, R., Carluccio, L.C., Carati, A. and Parker, W.O. *Microporous Mesoporous Mat.*, **46**, 191–201 (2001)
(5) Sumitani, K., Sakai, T., Yamasaki, Y. and Onodera, T. *E. Patent EP 51318* (1982)
(6) Rohrbaugh, W.J. *private communication*
(7) Arranz, M., Perez-Pariente, J., Wright, P.A., Slawin, A.M.Z., Blasco, T., Gomez-Hortiguela, L. and Cora, F. *Chem. Mater.*, **17**, 4374–4385 (2005)

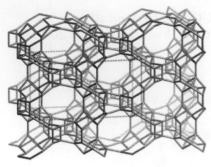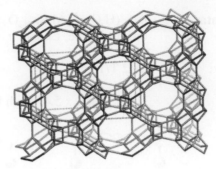

framework viewed along [100]

Idealized cell data: orthorhombic, *Imma*, $a = 10.2$Å, $b = 12.6$Å, $c = 21.7$Å

Coordination sequences and vertex symbols:

T_1 (16,1)	4	10	19	31	49	72	97	124	155	194	236	278	$4\cdot4\cdot6\cdot6\cdot6_2\cdot8$
T_2 (8,m..)	4	10	19	31	51	73	93	123	157	195	236	274	$4\cdot6_2\cdot4\cdot6_2\cdot6_2\cdot12_{18}$
T_3 (8,m..)	4	10	20	32	48	71	98	126	155	192	237	280	$4\cdot6\cdot4\cdot6\cdot6_2\cdot8$
T_4 (8,m..)	4	9	17	30	49	73	97	120	154	199	237	273	$4\cdot6\cdot4\cdot6\cdot4\cdot12_{26}$
T_5 (8,2..)	4	9	18	29	48	72	94	124	156	192	236	278	$4\cdot4\cdot4\cdot8\cdot6_2\cdot6_2$

Secondary building units: see *Compendium*

Composite building units:

lau

bph

Materials with this framework type:
    *EMM-3[1]

Crystal chemical data: [Al$_{24}$P$_{24}$O$_{96}$]-**EZT**
monoclinic, *I2/m*

$a = 10.3132$ Å, $b = 12.6975$ Å, $c = 21.8660$ Å, $\alpha = 89.656°$ [1]

Framework density: 16.8 T/1000Å^3

Channels: [100] **12** 6.5 x 7.4*

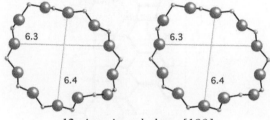

12-ring viewed along [100]

References:
(1)   Afeworki, M., Dorset, D.L., Kennedy, G.J. and Strohmaier, K.G. *Chem. Mater.*, **18**, 1697–1704 (2006)

FAR

Framework Type Data

$P6_3/mmc$

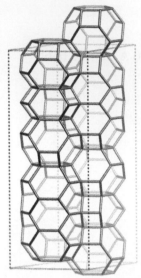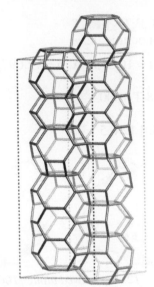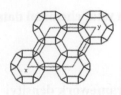

framework viewed normal to [001] (upper right: projection down [001])

Idealized cell data: hexagonal, $P6_3/mmc$, $a = 12.6$Å, $c = 35.7$Å

Coordination sequences and vertex symbols:

T_1 (24,1)	4	10	20	34	52	74	101	133	169	208	251	299	·4·4·6·6·6·6
T_2 (24,1)	4	10	20	34	53	76	102	132	166	206	251	299	4·4·6·6·6·6
T_3 (24,1)	4	10	20	34	53	76	102	132	166	205	248	294	4·6·4·6·6·6
T_4 (12,m..)	4	10	20	34	54	78	104	134	168	208	252	298	4·6·4·6·6·6

Secondary building units: 6

Framework description: ABCABABACBACAC sequence of 6-rings

Composite building units:

can	sod	lio

Materials with this framework type:
*Farneseite[1]

Farneseite   Type Material Data   **FAR**

Crystal chemical data: $|(Na,K)_{46}Ca_{10}(H_2O)_6(SO_4)_{12}|\,[Si_{42}Al_{42}O_{168}]$-**FAR**
hexagonal, $P6_3/m$, $a = 12.8784$Å, $c = 37.0078$Å [1]

Framework density: 15.8 T/1000Å^3

Channels: apertures formed by 6-rings only

References:
(1)   Cámara, F., Bellatreccia, F., Della Ventura, G. and Mottana, A. *Eur. J. Mineral.*, **17**, 839–846 (2005)

FAU

Framework Type Data

$Fd\overline{3}m$

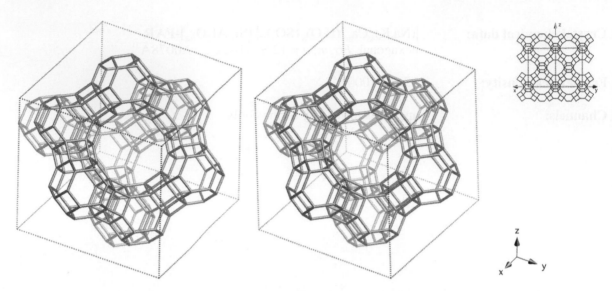

framework viewed along [111] (upper right: projection down [110])

Idealized cell data: cubic, $Fd\overline{3}m$ (origin choice 2), $a = 24.3$Å

Coordination sequences and vertex symbols:

   T_1 (192,1)   4   9   16   25   37   53   73   96   120   145   4·4·4·6·6·12

Secondary building units: 6-6 or 6-2 or 6 or 4-2 or 1-4-1 or 4

Composite building units:

   d6r   *sod*

Materials with this framework type:

*Faujasite[1,2]
[Al-Ge-O]-**FAU**[3,4]
[Co-Al-P-O]-**FAU**[5]
[Ga-Al-Si-O]-**FAU**[6]
[Ga-Ge-O]-**FAU**[3]
[Ga-Si-O]-**FAU**[7]
Beryllophosphate X[8,9]

Dehydrated Na-X[10]
Dehydrated US-Y[11]
LZ-210[12]
Li-LSX[13]
SAPO-37[14]
Siliceous Na-Y[15]
Zeolite X (Linde X)[16,17]

Zeolite Y (Linde Y)[18,19]
Zincophosphate X[9]
EMT-FAU structural intermediates:
   CSZ-1[20]
   ECR-30[21]
   ZSM-20[22]
   ZSM-3[23]

Faujasite

Type Material Data

FAU

Crystal chemical data: $|(Ca,MgNa_2)_{29} (H_2O)_{240}| [Al_{58}Si_{134}O_{384}]$-**FAU**
cubic, $Fd\overline{3}m$, $a = 24.74\text{Å}^{(2)}$

Framework density: $12.7 \text{ T}/1000\text{Å}^3$

Channels: <111> **12** 7.4 x 7.4***

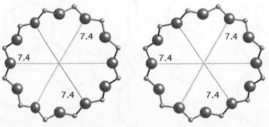

12-ring viewed along <111>

References:

(1) Bergerhoff, G., Baur, W.H. and Nowacki, W. *N. Jb. Miner. Mh.*, 193–200 (1958)
(2) Baur, W.H. *Am. Mineral.*, **49**, 697–704 (1964)
(3) Barrer, R.M., Baynham, J.W., Bultitude, F.W. and Meier, W.M. *J. Chem. Soc.*, 195–208 (1959)
(4) Johnson, G.M., Lee, Y.J., Tripathi, A. and Parise, J.B. *Microporous Mesoporous Mat.*, **31**, 195–204 (1999)
(5) Feng, P., Bu, X. and Stucky, G.D. *Nature*, **388**, 735–741 (1997)
(6) Occelli, M.L., Schweizer, A.E., Fild, C., Schwering, G., Eckert, H. and Auroux, A. *J. Catal.*, **192**, 119–127 (2000)
(7) Occelli, M.L., Schwering, G., Fild, C., Eckert, H., Auroux, A. and Iyer, P.S. *Microporous Mesoporous Mat.*, **34**, 15–22 (2000)
(8) Harrison, W.T.A., Gier, T.E., Moran, K.L., Nicol, J.M., Eckert, H. and Stucky, G.D. *Chem. Mater.*, **3**, 27–29 (1991)
(9) Gier, T.E. and Stucky, G.D. *Zeolites*, **12**, 770–775 (1992)
(10) Olson, D.H. *Zeolites*, **15**, 439–443 (1995)
(11) Parise, J.B., Corbin, D.R., Abrams, L. and Cox, D.E. *Acta Crystallogr.*, **C40**, 1493–1497 (1984)
(12) Breck, D.W. and Skeels, G.W. *U.S. Patent 4,503,023* (1985)
(13) Feuerstein, M. and Lobo, R.F. *Chem. Mater.*, 2197–2204 (1998)
(14) Lok, B.M., Messina, C.A., Patton, R.L., Gajek, R.T., Cannan, T.R. and Flanigen, E.M. *J. Am. Chem. Soc.*, **106**, 6092–6093 (1984)
(15) Hriljac, J.J., Eddy, M.M., Cheetham, A.K., Donohue, J.A. and Ray, G.J. *J. Solid State Chem.*, **106**, 66–72 (1993)
(16) Milton, R.M. *U.S. Patent 2,882,244* (1959)
(17) Olson, D.H. *J. Phys. Chem.*, **74**, 2758–2764 (1970)
(18) Breck, D.W. *U.S. Patent 3,130,007* (1964)
(19) Costenoble, M.L., Mortier, W.J. and Uytterhoeven, J.B. *J. Chem. Soc., Faraday Trans. I*, **72**, 1877–1883 (1976)
(20) Barrett, M.G. and Vaughan, D.E.W. *UK Patent GB 2,076,793* (1981)
(21) Vaughan, D.E.W. *E. Patent 0,351,461* (1989)
(22) Newsam, J.M., Treacy, M.M.J., Vaughan, D.E.W., Strohmaier, K.G. and Mortier, W.J. *Chem. Commun.*, 493–495 (1989)
(23) Kokotailo, G.T. and Ciric, J. *Adv. Chem. Ser.*, **101**, 109–121 (1971)

FER

Framework Type Data

Immm

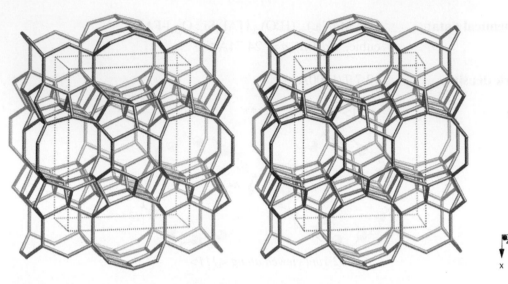

framework viewed along [001]

Idealized cell data: orthorhombic, *Immm*, $a = 19.0$Å, $b = 14.3$Å, $c = 7.5$Å

Coordination sequences and vertex symbols:

T_1 (16,1)	4	12	21	39	66	95	126	169	221	265	$5 \cdot 5 \cdot 5 \cdot 5_2 \cdot 5 \cdot 8$
T_2 (8,*m*)	4	12	27	43	62	97	139	172	206	264	$5 \cdot 8 \cdot 5 \cdot 8 \cdot 5_2 \cdot 6$
T_3 (8,*m*)	4	12	20	35	67	104	121	157	223	276	$5 \cdot 5_2 \cdot 5 \cdot 5_2 \cdot 10_2 \cdot 12_2$
T_4 (4,2*mm*)	4	12	23	40	66	96	131	164	214	272	$5 \cdot 5 \cdot 5 \cdot 5 \cdot 5_2 \cdot 6$

Secondary building units: 5-1

Composite building units:

fer

Materials with this framework type:

*Ferrierite[1]
[B-Si-O]-**FER**[2]
[Ga-Si-O]-**FER**[3]
[Si-O]-**FER**[4,5]

FU-9[6]
ISI-6[7]
Monoclinic ferrierite[8]
NU-23[9]

Sr-D[10]
ZSM-35[11]

Ferrierite Type Material Data **FER**

Crystal chemical data: |Mg$_2$Na$_2$ (H$_2$O)$_{18}$| [Al$_6$Si$_{30}$O$_{72}$]-**FER**
orthorhombic, *Immm*, a = 19.156Å, b = 14.127Å, c = 7.489Å [1]

Framework density: 17.8 T/1000Å^3

Channels: [001] **10** 4.2 x 5.4* ↔ [010] **8** 3.5 x 4.8*

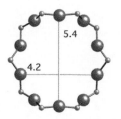 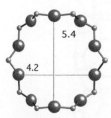 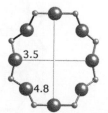 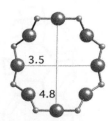

10-ring viewed along [001] *8-ring viewed along [010]*

References:
(1) Vaughan, P.A. *Acta Crystallogr.*, **21**, 983–990 (1966)
(2) Perego, G., Bellussi, G., Millini, R., Alberti, A. and Zanardi, S. *Microporous Mesoporous Mat.*, **56**, 193–202 (2002)
(3) Jacob, N.E., Joshi, P.N., Shaikh, A.A. and Shiralkar, V.P. *Zeolites*, **13**, 430–434 (1993)
(4) Gies, H. and Gunawardane, R.P. *Zeolites*, **7**, 442–445 (1987)
(5) Morris, R.E., Weigel, S.J., Henson, N.J., Bull, L.M., Janicke, M.T., Chmelka, B.F. and Cheetham, A.K. *J. Am. Chem. Soc.*, **116**, 11849–11855 (1994)
(6) Seddon, D. and Whittam, T.V. E. *Patent B-55,529* (1985)
(7) Morimoto, N., Takatsu, K. and Sugimoto, M. *U.S. Patent 4,578,259* (1986)
(8) Gramlich-Meier, R., Gramlich, V. and Meier, W.M. *Am. Mineral.*, **70**, 619–623 (1985)
(9) Whittam, T.V. E. *Patent A-103,981* (1984)
(10) Barrer, R.M. and Marshall, D.J. *J. Chem. Soc.*, 2296–2305 (1964)
(11) Plank, C.J., Rosinski, E.J. and Rubin, M.K. *U.S. Patent 4,016,245* (1977)

FRA

Framework Type Data

$P\bar{3}m1$

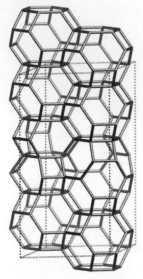
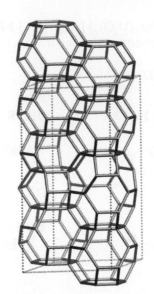
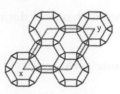

framework viewed normal to [001] (upper right: projection down [001])

Idealized cell data: trigonal, $P\bar{3}m1$, $a = 12.7$Å, $c = 25.3$Å

Coordination sequences and vertex symbols:

T_1 (12,1)	4	10	20	34	52	74	100	130	165	205	4·4·6·6·6·6
T_2 (12,1)	4	10	20	34	52	74	101	133	168	206	4·4·6·6·6·6
T_3 (12,1)	4	10	20	34	52	74	101	133	168	206	4·6·4·6·6·6
T_4 (12,1)	4	10	20	34	53	76	102	132	166	206	4·4·6·6·6·6
T_5 (6,2)	4	10	20	34	52	74	100	130	164	202	4·4·6·6·6·6
T_6 (6,2)	4	10	20	34	54	78	104	134	168	208	4·4·6·6·6·6

Secondary building units: 6 or 4

Framework description: ABCABACABC sequence of 6-rings

Composite building units:

can	*sod*	*los*

Materials with this framework type:
*Franzinite[1,2]

Franzinite

Type Material Data

FRA

Crystal chemical data: $|(Na,K)_{30} Ca_{10} (H_2O)_2 (SO_4)_{10}| [Al_{30}Si_{30}O_{120}]$-**FRA**
trigonal, $P321$, $a = 12.916$Å, $c = 26.543$Å [1]

Framework density: 15.6 T/1000Å^3

Channels: apertures formed by 6-rings only

References:
(1) Ballirano, P., Bonaccorsi, E., Maras, A. and Merlino, S. *Can. Mineral.*, **38**, 657–668 (2000)
(2) Ballirano, P. and Maras, A. *Powder Diffraction*, **16**, 216–219 (2001)

GIS

Framework Type Data

I4₁/amd

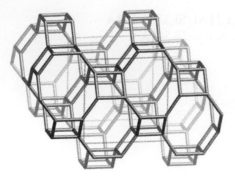 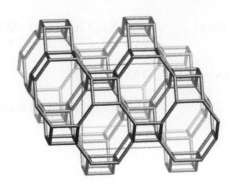

framework viewed along [100]

Idealized cell data: tetragonal, *I4₁/amd* (origin choice 2), $a = 9.8Å$, $c = 10.2Å$

Coordination sequences and vertex symbols:

T₁ (16,2)	4	9	18	32	48	67	92	120	150	185	4·4·4·8₂·8·8

Secondary building units: 8 or 4

Composite building units:

dcc   *gis*

double crankshaft chain

Materials with this framework type:

*Gismondine[1]
[Al-Co-P-O]-**GIS**[2]
[Al-Ge-O]-**GIS**[3,4]
[Al-P-O]-**GIS**[5]
[Be-P-O]-**GIS**[6]
[Co-Al-P-O]-**GIS**[7]
[Co-Ga-P-O]-**GIS**[8]
[Co-P-O]-**GIS**[9]
[Ga-Si-O]-**GIS**[10]
[Mg-Al-P-O]-**GIS**[7]
[Zn-Al-As-O]-**GIS**[11]
[Zn-Co-B-P-O]-**GIS**[12]

[Zn-Ga-As-O]-**GIS**[11]
[Zn-Ga-P-O]-**GIS**[13]
|(C₃H₁₂N₂)₄|[Be₈P₈O₃₂]-**GIS**[14]
|(C₃H₁₂N₂)₄|[Zn₈P₈O₃₂]-**GIS**[15]
|(NH₄)₄|[Zn₄B₄P₈O₃₂]-**GIS**[16]
|Cs₄|[Zn₄B₄P₈O₃₂]-**GIS**[16]
|Rb₄|[Zn₄B₄P₈O₃₂]-**GIS**[16]
Amicite[17]
Garronite[17,19]
Gobbinsite[20]
High-silica Na-P[21]
Low-silica Na-P (MAP)[22]

MAPO-43[23]
MAPSO-43[24,25]
Na-P1[26]
Na-P2[27]
SAPO-43[28]
Synthetic Ca-garronite[29]
Synthetic amicite[30]
Synthetic garronite[30]
Synthetic gobbinsite[30]
TMA-gismondine[31]

Gismondine   Type Material Data   GIS

Crystal chemical data: |Ca$_4$ (H$_2$O)$_{16}$| [Al$_8$Si$_8$O$_{32}$]-**GIS**
monoclinic, $P2_1/a$

a = 9.843Å, b = 10.023Å, c = 10.616Å, γ = 92.417° [1]
(Relationship to unit cell of Framework Type: a' = a, b' = b, c' = c)

Framework density: 15.3 T/1000Å^3

Channels: {[100] **8** 3.1 x 4.5 ↔ [010] **8** 2.8 x 4.8}***
(variable due to considerable flexibility of framework)
see Appendix A for 8-rings viewed along [100] and [010]

References:
(1) Fischer, K. and Schramm, V. *Adv. Chem. Ser.*, **101**, 250–258 (1971)
(2) Feng, P., Bu, X. and Stucky, G.D. *Nature*, **388**, 735–741 (1997)
(3) Johnson, G.M., Tripathi, A. and Parise, J.B. *Chem. Mater.*, **11**, 10+ (1999)
(4) Tripathi, A., Parise, J.B., Kim, S.J., Lee, Y., Johnson, G.M. and Uh, Y.S. *Chem. Mater.*, **12**, 3760–3769 (2000)
(5) Paillaud, J.L., Marler, B. and Kessler, H. *Chem. Commun.*, 1293–1294 (1996)
(6) Zhang, H., Chen, M., Shi, Z., Bu, X., Zhou, Y., Xu, X. and Zhao, D. *Chem. Mater.*, **13**, 2042–2048 (2001)
(7) Feng, P., Bu, X., Gier, T.E. and Stucky, G.D. *Microporous Mesoporous Mat.*, **23**, 221–229 (1998)
(8) Cowley, A.R. and Chippindale, A.M. *Chem. Commun.*, 673–674 (1996)
(9) Yuan, H.M., Chen, J.S., Zhu, G.S.,Li, J.Y., Yu, J.H., Yang, G.D. and Xu, R. *Inorg. Chem.*, **39**, 1476–1479 (2000)
(10) Cho, H.H., Kim, S.H., Kim, Y.G., Kim, Y.C., Koller, H., Camblor, M.A. and Hong, S.B. *Chem. Mater.*, **12**, 2292–2300 (2000)
(11) Feng, P., Zhang, T. and Bu, X. *J. Am. Chem. Soc.*, **123**, 8608–8609 (2001)
(12) Schafer, G., Borrmann, H. and Kniep, R. *Microporous Mesoporous Mat.*, **41**, 161–167 (2000)
(13) Chippindale, A.M., Cowley, A.R. and Peacock, K.J. *Microporous Mesoporous Mat.*, **24**, 133–141 (1998)
(14) Harrison, W.T.A. *Acta Crystallogr.*, **C57**, 891–892 (2001)
(15) Harrison, W.T.A. *International Journal of Inorganic Materials*, **3**, 179–182 (2001)
(16) Kniep, R., Schäfer, G., Engelhardt, H. and Boy, I. *Angew. Chem. Int. Ed.*, **38**, 3642–3644 (1999)
(17) Alberti, A. and Vezzalini, G. *Acta Crystallogr.*, **B35**, 2866–2869 (1979)
(18) Artioli, G. *Am. Mineral.*, **77**, 189–196 (1992)
(19) Artioli, G. and Marchi, M. *Powder Diffraction*, **14**, 190–194 (1999)
(20) McCusker, L.B., Baerlocher, Ch. and Nawaz, R. *Z. Kristallogr.*, **171**, 281–289 (1985)
(21) Håkansson, U., Fälth, L. and Hansen, S. *Acta Crystallogr.*, **C46**, 1363–1364 (1990)
(22) Albert, B.R., Cheetham, A.K., Stuart, J.A. and Adams, C.J.` *Microporous Mesoporous Mat.*, **21**, 133–142 (1998)
(23) Pluth, J.J., Smith, J.V. and Bennett, J.M. *J. Am. Chem. Soc.*, **111**, 1692–1698 (1989)
(24) Flanigen, E.M., Lok, B.M., Patton, R.L. and Wilson, S.T. *Pure Appl. Chem.*, **58**, 1351–1358 (1986)
(25) Flanigen, E.M., Lok, B.M., Patton, R.L. and Wilson, S.T. *Proc. 7th Int. Zeolite Conf.*, pp. 103–112 (1986)
(26) Baerlocher, Ch. and Meier, W.M. *Z. Kristallogr.*, **135**, 339–354 (1972)
(27) Hansen, S., Håkansson, U. and Fälth, L. *Acta Crystallogr.*, **C46**, 1361–1362 (1990)
(28) Helliwell, M., Kaucic, V., Cheetham, G.M.T., Harding, M.M., Kariuki, B.M. and Rizkallah, P.J. *Acta Crystallogr.*, **B49**, 413–420 (1993)
(29) Schropfer, L and Joswig, W. *Eur. J. Mineral.*, **9**, 53–65 (1997)
(30) Ghobarkar, H. and Schaef, O. *Mater. Res. Bull.*, **34**, 517–525 (1999)
(31) Baerlocher, Ch. and Meier, W.M. *Helv. Chim. Acta*, **53**, 1285–1293 (1970)

GIU

Framework Type Data

$P6_3/mmc$

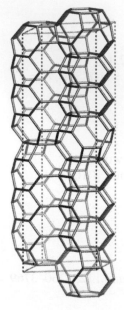
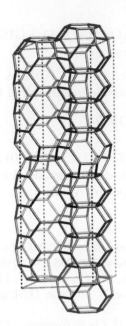
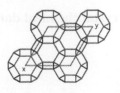

framework viewed normal to [001] (upper right: projection down [001])

Idealized cell data: hexagonal, $P6_3/mmc$, $a = 12.6$Å, $c = 41.0$Å

Coordination sequences and vertex symbols:

T_1 (24,1)	4	10	20	34	53	76	102	132	167	209	255	301		4·4·6·6·6·6
T_2 (24,1)	4	10	20	34	53	76	102	132	166	206	251	299		4·6·4·6·6·6
T_3 (24,1)	4	10	20	34	54	78	104	134	168	209	254	300		4·6·4·6·6·6
T_4 (12,m..)	4	10	20	34	54	78	104	134	168	208	252	298		4·6·4·6·6·6
T_5 (12,.2.)	4	10	20	34	52	74	102	136	172	208	248	298		4·4·6·6·6·6

Secondary building units: 6 or 4

Framework description: ABABABACBABABABC sequence of 6-rings

Composite building units:

can   sod

Materials with this framework type:
*Giuseppettite[1]

Giuseppettite
Type Material Data
GIU

Crystal chemical data: $|Na_{42}K_{16}Ca_6 (SO_4)_{10}Cl_2) (H_2O)_5| [Si_{48}Al_{48}O_{192}]$-**GIU**
trigonal, $P31c$, $a = 12.856$Å, $c = 42.256$Å[1]

Framework density: 15.9 T/1000Å^3

Channels: apertures formed by 6-rings only

References:
(1) Bonaccorsi, E. *Microporous Mesoporous Mat.*, **73**, 129–136 (2004)

GME

Framework Type Data

$P6_3/mmc$

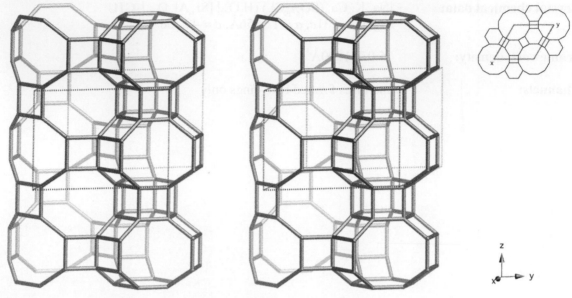

framework viewed normal to [001] (upper right: projection down [001])

Idealized cell data: hexagonal, $P6_3/mmc$, $a = 13.7$Å, $c = 9.9$Å

Coordination sequences and vertex symbols:

T_1 (24,1)	4	9	17	29	45	65	89	116	144	175

4·4·4·8·6·8

Secondary building units: 12 or 6-6 or 8 or 6 or 4-2 or 4

Framework description: AABB sequence of 6-rings

Composite building units:

dcc	d6r	gme
double crankshaft chain		

Materials with this framework type:
*Gmelinite[1]
[Be-P-O]-**GME**[2]
K-rich gmelinite[3]
Synthetic fault-free gmelinite[4]

Gmelinite

Type Material Data

GME

Crystal chemical data: |(Ca,Na$_2$)$_4$ (H$_2$O)$_{24}$| [Al$_8$Si$_{16}$O$_{48}$]-**GME**
hexagonal, $P6_3/mmc$, $a = 13.75$Å, $c = 10.05$Å $^{(1)}$

Framework density: 14.6 T/1000Å^3

Channels: [001] **12** 7.0 x 7.0* ↔ ⊥ [001] **8** 3.6 x 3.9**

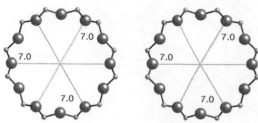 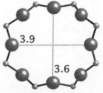

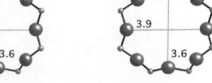

12-ring viewed along [001]   *8-ring viewed normal to [001]*

References:

(1) Fischer, K. *N. Jb. Miner. Mh.*, 1-13 (1966)
(2) Zhang, H., Chen, M., Shi, Z., Bu, X., Zhou, Y., Xu, X. and Zhao, D. *Chem. Mater.*, **13**, 2042–2048 (2001)
(3) Vezzalini, G., Quartieri, S. and Passaglia, E. *N. Jb. Miner. Mh.*, 504–516 (1990)
(4) Daniels, R.H., Kerr, G.T. and Rollmann, L.D. *J. Am. Chem. Soc.*, **100**, 3097–3100 (1978)

GON

Framework Type Data

Cmmm

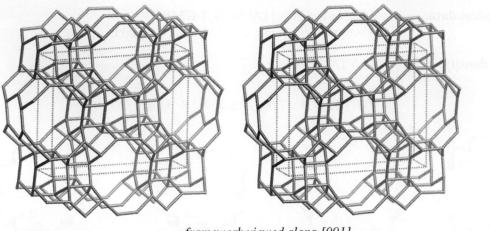

framework viewed along [001]

Idealized cell data: orthorhombic, *Cmmm*, $a = 16.9$Å, $b = 20.4$Å, $c = 5.3$Å

Coordination sequences and vertex symbols:

T_1 (8,*m*)	4	12	21	39	63	85	117	154	192	242	$5·6·5·6·5_2·6$
T_2 (8,*m*)	4	12	25	38	57	86	119	158	194	233	$5·6·5·6·6·12_6$
T_3 (8,*m*)	4	11	22	38	58	86	121	156	191	229	$4·6_2·5·6·5·6$
T_4 (8,*m*)	4	11	22	38	63	91	115	147	195	244	$4·6_2·5·6·5·6$

Secondary building units: 5-3

Composite building units:

mtw   *cas*

Materials with this framework type:
  *GUS-1[(1)]

GUS-1 **Type Material Data** **GON**

Crystal chemical data: [Si$_{32}$O$_{64}$]-**GON**
orthorhombic, *C*222, *a* = 16.421Å, *b* = 20.054Å, *c* = 5.046Å [1]

Framework density: 19.3 T/1000Å^3

Channels: [001] **12** 5.4 x 6.8*

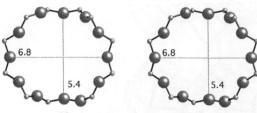
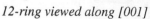
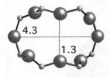
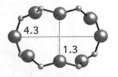

12-ring viewed along [001] *8-ring viewed along [001]*

References:
(1) Plévert, J., Kubota, Y., Honda, T., Okubo, T. and Sugi, Y. *Chem. Commun.*, 2363–2364 (2000)

GOO

Framework Type Data

*C*222₁

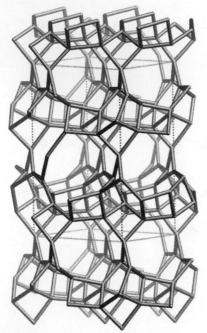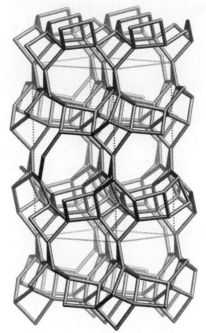

Idealized cell data: orthorhombic, $C222_1$, $a = 8.7$Å, $b = 11.0$Å, $c = 17.5$Å

Coordination sequences and vertex symbols:

T_1 (8,1)	4	9	19	36	56	80	102	132	180	220	$4 \cdot 4 \cdot 4 \cdot 8_5 \cdot 6_2 \cdot 8_2$
T_2 (8,1)	4	10	20	36	55	77	108	140	174	219	$4 \cdot 4 \cdot 6 \cdot 8_3 \cdot 6_3 \cdot 8_2$
T_3 (8,1)	4	9	19	36	56	76	106	142	172	217	$4 \cdot 4 \cdot 4 \cdot 8_2 \cdot 6 \cdot 8_2$
T_4 (4,2)	4	10	20	34	58	82	102	136	176	220	$4 \cdot 4 \cdot 6_4 \cdot 8_4 \cdot 8_2 \cdot 8_2$
T_5 (4,2)	4	10	20	34	52	78	110	140	176	212	$4 \cdot 4 \cdot 8_2 \cdot 10_6 \cdot 8_3 \cdot 8_3$

Secondary building units: see *Compendium*

Materials with this framework type:
   *Goosecreekite[1]

Goosecreekite

Type Material Data

GOO

Crystal chemical data: $|Ca_2 (H_2O)_{10}| [Al_4Si_{12}O_{32}]$-**GOO**
monoclinic, $P2_1$

$a = 7.401$Å, $b = 17.439$Å, $c = 7.293$Å, $\beta = 105.44°$ [1]
(Relationship to unit cell of Framework Type:

$a' = a/(2\cos(\beta'/2))$, $b' = c$, $c' = a/(2\cos(\beta'/2))$
or, as vectors, $\mathbf{a'} = (\mathbf{a} + \mathbf{b})/2$, $\mathbf{b'} = \mathbf{c}$, $\mathbf{c'} = (\mathbf{a} - \mathbf{b})/2$

Framework density: 17.6 T/1000Å^3

Channels: [100] **8** 2.8 x 4.0* $\leftrightarrow$ [010] **8** 2.7 x 4.1* $\leftrightarrow$ [001] **8** 2.9 x 4.7*

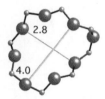 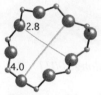   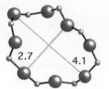 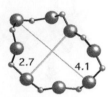

8-ring viewed along [100]   *8-ring viewed along [010]*

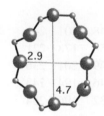 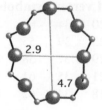

8-ring viewed along [001]

References:
(1) Rouse, R.C. and Peacor, D.R. *Am. Mineral.*, **71**, 1494–1501 (1986)

HEU

Framework Type Data

C2/m

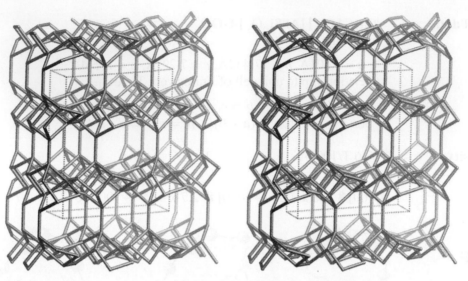

framework viewed along [001]

Idealized cell data: monoclinic, $C2/m$, $a = 17.5$Å, $b = 17.6$Å, $c = 7.4$Å, $\beta = 116.1°$

Coordination sequences and vertex symbols:

T_1 (8,1)	4	10	20	34	62	85	104	148	201	241	$4·5·4·5·5·8$
T_2 (8,1)	4	11	23	39	55	82	127	158	178	221	$4·8·5·8·5·8$
T_3 (8,1)	4	10	19	37	58	84	109	149	201	236	$4·5·4·8·5·5$
T_4 (8,1)	4	11	21	35	61	89	111	146	194	243	$4·5·5·5·5·8$
T_5 (4,2)	4	12	18	34	62	88	110	132	196	254	$5·5·5_2·5_2·10·10$

Secondary building units: 4-4=1

Composite building units:
  bre

Materials with this framework type:
  *Heulandite[1,2]
  Clinoptilolite[3]
  Dehyd. Ca,NH$_4$-Heulandite[4]

  Heulandite-Ba[5]
  LZ-219[6]

Heulandite   Type Material Data   **HEU**

Crystal chemical data: |Ca$_4$ (H$_2$O)$_{24}$| [Al$_8$Si$_{28}$O$_{72}$]-**HEU**
monoclinic, *Cm*

$$a = 17.718\text{Å}, \ b = 17.897\text{Å}, \ c = 7.428\text{Å}, \ \beta = 116.42°\ ^{(2)}$$

Framework density: 17.1 T/1000Å^3

Channels: {[001] **10** 3.1 x 7.5* + **8** 3.6 x 4.6*} ↔ [100] **8** 2.8x 4.7*
(variable due to considerable flexibility of framework)

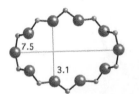 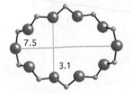 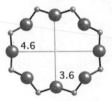 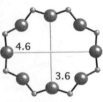

10-ring viewed along [001]   *8-ring, also along [001]*

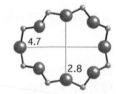 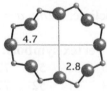

8-ring viewed along [100]

References:
(1) Merkle, A.B. and Slaughter, M. *Am. Mineral.*, **52**, 273–276 (1967)
(2) Alberti, A. *Tschermaks Min. Petr. Mitt.*, **18**, 129–146 (1972)
(3) Koyama, K. and Takeuchi, Y. *Z. Kristallogr.*, **145**, 216–239 (1977)
(4) Mortier, W.J. and Pearce, J.R. *Am. Mineral.*, **66**, 309–314 (1981)
(5) Larsen, A.O., Nordrum, F.S., Dobelin, N., Armbruster, T., Petersen, O.V. and Erambert, M. *Eur. J. Mineral.*, **17**, 143–153 (2005)
(6) Breck, D.W. and Skeels, G.W. *U.S. Patent 4,503,023* (1985)

IFR

Framework Type Data

C2/m

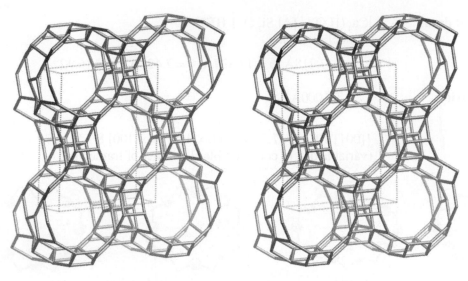

framework viewed along [001]

Idealized cell data: monoclinic, *C2/m*, $a = 18.6$Å, $b = 13.4$Å, $c = 7.6$Å, $\beta = 102.3°$

Coordination sequences and vertex symbols:

T_1 (8,1)	4	11	20	32	53	77	100	135	166	199	$4 \cdot 6_2 \cdot 5 \cdot 6 \cdot 6 \cdot 12_3$
T_2 (8,1)	4	10	18	31	52	77	103	127	159	210	$4 \cdot 5 \cdot 4 \cdot 6 \cdot 5 \cdot 6_2$
T_3 (8,1)	4	10	19	32	53	78	102	126	162	209	$4 \cdot 6 \cdot 4 \cdot 6 \cdot 5 \cdot 6_2$
T_4 (8,1)	4	10	21	35	50	74	105	133	165	206	$4 \cdot 4 \cdot 5 \cdot 12_5 \cdot 6 \cdot 6_2$

Secondary building units: 6-2

Composite building units:

     bea   *lau*

Materials with this framework type:
  *ITQ-4[1,2]
  MCM-58[3]
  SSZ-42[4,5]

ITQ-4   **Type Material Data**   **IFR**

Crystal chemical data:	$[Si_{32}O_{64}]$-**IFR**
	monoclinic, $I2/m$

$a = 18.652\text{Å}, b = 13.496\text{Å}, c = 7.631\text{Å}, \beta = 101.98°$ [1]
(Relationship to unit cell of Framework Type:
as vectors, $\mathbf{a'} = \mathbf{a} + \mathbf{c}$, $\mathbf{b'} = \mathbf{b}$, $\mathbf{c'} = \mathbf{-c}$)

Framework density: 17 T/1000Å^3

Channels: [001] **12** 6.2 x 7.2*

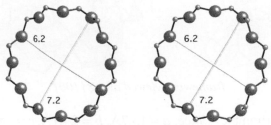

12-ring viewed along [001]

References:

(1) Barrett, P.A., Camblor, M.A., Corma, A., Jones, R.H. and Villaescusa, L.A. *Chem. Mater.*, **9**, 1713–1715 (1997)

(2) Barrett, P.A., Camblor, M.A., Corma, A., Jones, R.H. and Villaescusa, L.A. *J. Phys. Chem. B*, **102**, 4147–4155 (1998)

(3) Valyocsik, E.W. *WOP 9511196* (1995)

(4) Chen, C.Y., Finger, L.W., Medrud, R.C., Crozier, P.A., Chan, I.Y., Harris, T.V. and Zones, S.I. *Chem. Commun.*, 1775–1776 (1997)

(5) Chen, C.Y., Finger, L.W., Medrud, R.C., Kibby, C.L., Crozier, P.A., Chan, I.Y., Harris, T.V., Beck, L.W. and Zones, S.I. *Chem. Eur. Journal*, **4**, 1312–1323 (1998)

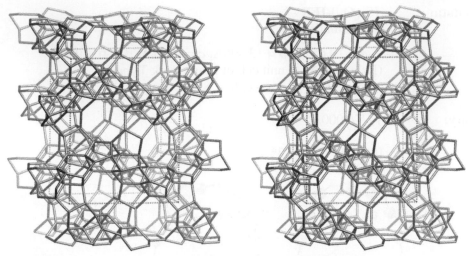

framework viewed along [100]

Idealized cell data: orthorhombic, *Cmce*, $a = 13.7$Å, $b = 24.1$Å, $c = 18.3$Å

Coordination sequences and vertex symbols:

T_1 (16,1)	4	11	24	41	69	95	122	165	221	276	323	374		$4 \cdot 8 \cdot 5 \cdot 5 \cdot 5 \cdot 5$
T_2 (16,1)	4	12	23	40	67	96	130	161	208	273	329	373		$5 \cdot 5 \cdot 5 \cdot 5 \cdot 5 \cdot 8$
T_3 (16,1)	4	12	22	40	62	96	127	166	213	258	320	381		$5 \cdot 5 \cdot 5 \cdot 5 \cdot 5 \cdot 6$
T_4 (16,1)	4	12	24	41	65	94	129	171	216	264	322	382		$5 \cdot 5 \cdot 5 \cdot 6_2 \cdot 5 \cdot 8$
T_5 (16,1)	4	12	25	40	63	94	134	177	208	258	324	393		$5 \cdot 6 \cdot 5_2 \cdot 8 \cdot 6 \cdot 6_2$
T_6 (8,*m*..)	4	12	22	37	65	95	129	170	212	259	310	397		$5 \cdot 5 \cdot 5 \cdot 5 \cdot 5 \cdot 6_2$
T_7 (8,*m*..)	4	12	24	38	62	97	137	168	211	260	312	389		$5 \cdot 5 \cdot 5 \cdot 5 \cdot 12 \cdot *$
T_8 (8,*m*..)	4	12	22	37	65	99	128	164	210	262	319	383		$5 \cdot 5 \cdot 5 \cdot 5 \cdot 5 \cdot 6_2$
T_9 (8,*m*..)	4	12	24	36	62	95	132	169	204	251	326	399		$5 \cdot 5 \cdot 5 \cdot 5 \cdot 12_6 \cdot *$

Secondary building units: see *Compendium*

Composite building units:

     cas   *non*   *ton*

Materials with this framework type:
    *ITQ-32[1]

ITQ-32   **Type Material Data**   **IHW**

Crystal chemical data: [Si$_{112}$O$_{224}$]-**IHW**
                               orthorhombic, *Cmce*, $a = 13.6988$Å, $b = 24.0665$Å, $c = 18.1968$Å [1]

Framework density: 18.7 T/1000Å^3

Channels: [100] **8** 3.5 x 4.3**
                               (There are 12-ring connections between the channels along [100] but
                               there is no continuous 12-ring channel.)

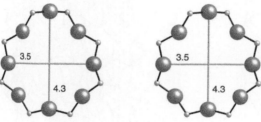

8-ring viewed along [100]

References:
(1) Cantin, A., Corma, A., Leiva, S., Rey, F., Rius, J. and Valencia, S. *J. Am. Chem. Soc.*, **127**, 11560–11561
     (2005)

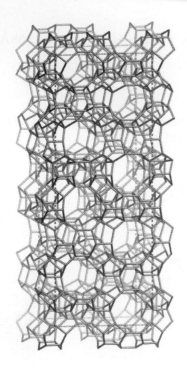 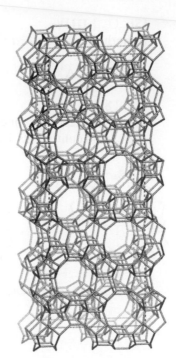

framework viewed along [100]

Idealized cell data: orthorhombic, *Cmcm*, $a = 14.3$Å, $b = 56.8$Å, $c = 20.3$Å

Coordination sequences and vertex symbols:
    see Appendix A for a list of the coordination sequences and vertex symbols for the 24 T-atoms

Secondary building units: 5-1

Composite building units:

mor	*mtt*	*ats*	*stf*	*fer*

Materials with this framework type:
*IM-5[1]

IM-5   **Type Material Data**   **IMF**

Crystal chemical data: $[Si_{288}O_{576}]$-**IMF**
orthorhombic, *Cmcm*, $a = 14.2088$Å, $b = 57.2368$Å, $c = 19.9940$Å [1]

Framework density: 17.7 T/1000Å^3

Channels: {[001] **10** 5.5 x 5.6 ↔ [100] **10** 5.3 x 5.4}** ↔ {[010] **10** 5.3 x 5.9}

↔ {[001] **10** 4.8 x 5.4 ↔ [100] **10** 5.1 x 5.3}**
(There is central 2D channel system (above: left) connected through
10-rings along [010] (above: after second ↔) to another 2D channel
system (above: after third ↔) on either side. There is no further
connection along [010].)

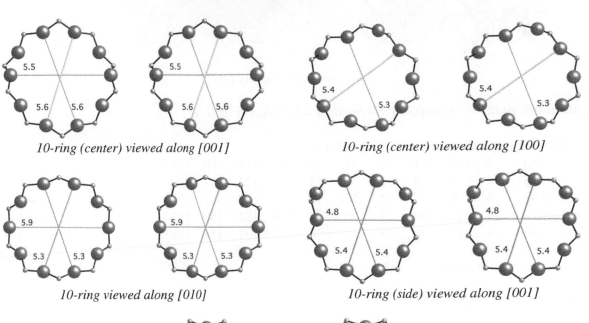

10-ring (center) viewed along [001]   *10-ring (center) viewed along [100]*

10-ring viewed along [010]   *10-ring (side) viewed along [001]*

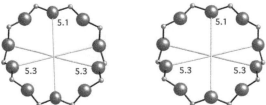

10-ring (side) viewed along [100]

References:
(1)   Baerlocher, Ch., Gramm, F., Massüger, L., McCusker, L.B., He, Z., Hovmöller, S. and Zou, X. *Science*, **315**, 1113–1116 (2007)

ISV

Framework Type Data

P4₂/mmc

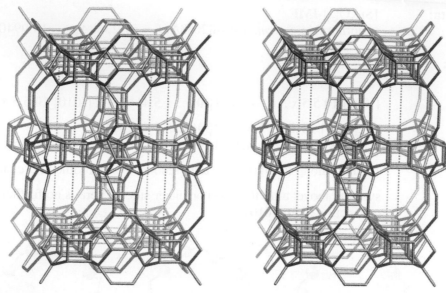

framework viewed along [100]

Idealized cell data: tetragonal, $P4_2/mmc$, $a = 12.9$Å, $c = 25.7$Å

Coordination sequences and vertex symbols:

T_1 (16,1)	4	12	17	30	48	72	99	128	160	199	$5\cdot5\cdot5_2\cdot12_5\cdot6\cdot6$
T_2 (16,1)	4	9	18	32	50	71	96	129	167	200	$4\cdot5\cdot4\cdot6\cdot4\cdot12_7$
T_3 (16,1)	4	9	18	32	50	72	97	128	167	203	$4\cdot5\cdot4\cdot6\cdot4\cdot12_4$
T_4 (8,m)	4	11	20	28	42	74	110	132	150	195	$4\cdot5_2\cdot5\cdot6\cdot5\cdot6$
T_5 (8,m)	4	11	20	28	41	70	105	131	154	188	$4\cdot5_2\cdot5\cdot6\cdot5\cdot6$

Secondary building units: 6-2

Composite building units:

d4r   *mor*   *mtw*

Materials with this framework type:
*ITQ-7[1]
[Ge-Si-O]-**ISV**[2]
|(C₁₅H₂₉N)₄F₄|[Si₆₄O₁₂₈]-**ISV**[3]
|BCHP|[Si$_x$Al$_y$Ge$_{II}$O₁₂₈]-**ISV**[4]

Crystal chemical data: [Si$_{64}$O$_{128}$]-**ISV**
tetragonal, $P4_2/mmc$, a = 12.853 Å, c = 25.214Å [1]

Framework density: 15.4 T/1000Å^3

Channels: <100> **12** 6.1 x 6.5** ↔ [001] **12** 5.9 x 6.6*

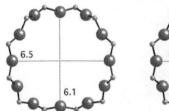 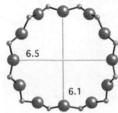 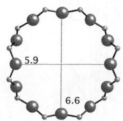 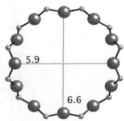

12-ring viewed along <100>   *12-ring viewed along [001]*

References:
(1) Villaescusa, L.A., Barrett, P.A. and Camblor, M.A. *Angew. Chem., Int. Ed.*, **38**, 1997–2000 (1999)
(2) Blasco, T., Corma, A., DiazCabanas, M.J., Rey, F., VidalMoya, J.A. and ZicovichWilson, C.M. *J. Phys. Chem. B*, **106**, 2634–2642 (2002)
(3) Song, J.Q., Marler, B. and Gies, H. *Compt. Rend. Chimie*, **8**, 341–352 (2005)
(4) Leiva, S., Sabater, M.J., Valencia, S., Sastre, G., Fornes, V., Rey, F. and Corma, A. *Compt. Rend. Chimie*, **8**, 369–378 (2005)

ITE

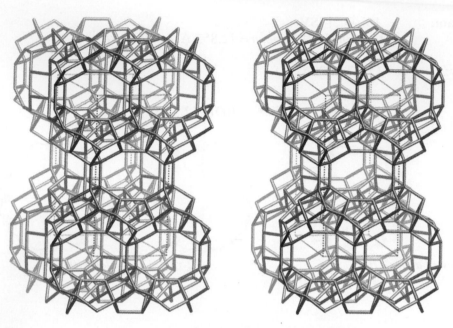

framework viewed along [001]

Idealized cell data: orthorhombic, *Cmcm*, $a = 20.8$Å, $b = 9.8$Å, $c = 20.0$Å

Coordination sequences and vertex symbols:

T_1 (16,1)	4	11	21	34	53	78	108	138	168	211	4·6·5·6·5·8
T_2 (16,1)	4	10	21	36	54	77	102	135	181	217	4·4·5·8·5·8
T_3 (16,1)	4	10	19	31	50	82	107	132	168	209	4·5·4·6·5·5
T_4 (16,1)	4	10	18	31	55	77	105	136	166	216	4·5·4·8·5·5

Secondary building units: 4

Composite building units:
   rth

Materials with this framework type:
   *ITQ-3[1]
   Mu-14[2]
   SSZ-36 (**ITE-RTH** structural intermediate)[3]

ITQ-3   Type Material Data   ITE

Crystal chemical data: $[Si_{64}O_{128}]$-**ITE**
orthorhombic, *Cmcm*, *a* = 20.622Å, *b* = 9.724Å, *c* = 19.623Å [1]

Framework density: 16.3 T/1000Å^3

Channels: [010] **8** 3.8 x 4.3* ↔ [001] **8** 2.7 x 5.8*

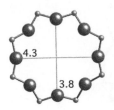 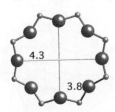 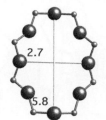 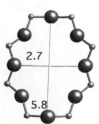

       8-ring viewed along [010]   *8-ring viewed along [001]*

References:
(1) Camblor, M.A., Corma, A., Lightfoot, P., Villaescusa, L.A. and Wright, P.A. *Angew. Chem., Int. Ed.*, **36**, 2659–2661 (1997)
(2) Valtchev, V., Paillaud, J.L., Lefebvre, T., LeNouen, D. and Kessler, H. *Microporous Mesoporous Mat.*, **38**, 177–185 (2000)
(3) Wagner, P., Nakagawa, Y., Lee, G.S., Davis, M.E., Elomari, S., Medrud, R.C. and Zones, S.I. *J. Am. Chem. Soc.*, **122**, 263–273 (2000)

ITH

Framework Type Data

Amm2

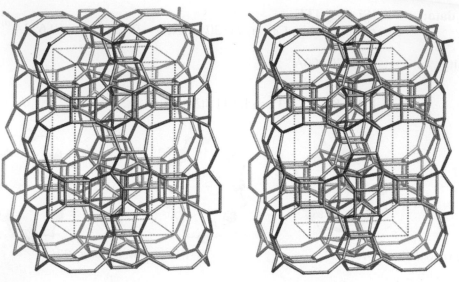

framework viewed along [010]

Idealized cell data: orthorhombic, *Amm2*, $a = 12.6$Å, $b = 11.7$Å, $c = 21.9$Å

Coordination sequences and vertex symbols:

T_1 (8,1)	4	9	18	35	61	90	116	146	190	244	296	338	$4 \cdot 6_2 \cdot 4 \cdot 6_2 \cdot 4 \cdot 10_2$
T_2 (8,1)	4	12	22	37	56	84	116	154	193	233	288	351	$5 \cdot 6 \cdot 5 \cdot 6 \cdot 6_2 \cdot 9_2$
T_3 (8,1)	4	9	19	37	60	86	117	151	189	238	296	349	$4 \cdot 6 \cdot 4 \cdot 6_2 \cdot 4 \cdot 10_2$
T_4 (8,1)	4	12	21	36	56	83	112	155	196	237	283	349	$5 \cdot 6 \cdot 5 \cdot 6_2 \cdot 5 \cdot 9$
T_5 (8,1)	4	11	22	39	57	80	114	156	202	239	283	343	$4 \cdot 9_2 \cdot 5 \cdot 5 \cdot 5 \cdot 6_2$
T_6 (4,m..)	4	11	22	32	51	87	116	139	183	241	296	343	$4 \cdot 5_2 \cdot 6 \cdot 6_2 \cdot 6 \cdot 6_2$
T_7 (4,m..)	4	11	20	32	55	85	112	141	185	243	294	339	$4 \cdot 5_2 \cdot 6_2 \cdot 6_2 \cdot 6_2 \cdot 6_2$
T_8 (4,.m.)	4	11	22	36	53	83	119	155	186	236	299	359	$4 \cdot 9_2 \cdot 5 \cdot 6_2 \cdot 5 \cdot 6_2$
T_9 (4,.m.)	4	11	20	34	57	81	117	155	192	234	297	357	$4 \cdot 5 \cdot 5 \cdot 6_2 \cdot 5 \cdot 6_2$

Secondary building units: see *Compendium*

Composite building units:

    d4r   *lau*   *mel*

Materials with this framework type:

  *ITQ-13[1,2]   Al-ITQ 13[3]   IM-7[4]

ITQ-13   **Type Material Data**   **ITH**

Crystal chemical data: $|((CH_3)_3N(CH_2)_6N(CH_3)_3)_2 \ F_4| \ [Si_{56}O_{112}]$**-ITH**
orthorhombic, $Amm2$, $a = 12.525$Å, $b = 11.391$Å, $c = 22.053$Å [2]

Framework density: 17.8 T/1000Å^3

Channels: [001] **10**   4.8 x 5.3* ↔ [010] **10**   4.8 x 5.1* ↔ [100] **9** 4.0 x 4.8*

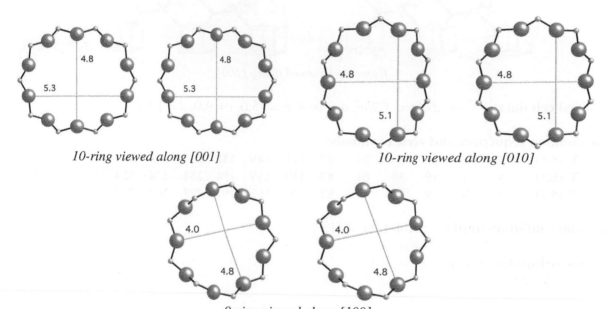

10-ring viewed along [001]   *10-ring viewed along [010]*

9-ring viewed along [100]

References:
(1)   Boix, T., Puche, M., Camblor, M.A. and Corma, A. *U.S. Patent 6,471,941 B1* (2002)
(2)   Corma, A., Puche, M., Rey, F., Sankar, G. and Teat, S.J. *Angew. Chem., Int. Ed.*, **42**, 1156–1159 (2003)
(3)   Castaneda, R., Corma, A., Fornes, V., Martinez-Triguero, J. and Valencia, S. *J. Catal.*, **238**, 79–87 (2006)
(4)   Bats, N., Rouleau, L., Paillaud, J.-L., Caullet, P., Mathieu, Y. and Lacombe, S. *Stud. Surf. Sci. Catal.*, **154**, 283–288 (2004)

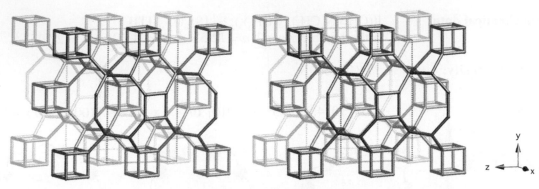

framework viewed along [100]

Idealized cell data: monoclinic, *C2/m*, *a* = 10.4, *b* = 15.0, *c* = 9.0, β = 105.6

Coordination sequences and vertex symbols:

T_1 (8,1)	4	12	23	36	56	87	121	149	181	231	288	337	$5·5·5·6·8·8$
T_2 (8,1)	4	9	19	38	61	83	110	149	194	234	274	329	$4·5·4·6·4·8_2$
T_3 (8,1)	4	9	19	38	62	82	108	152	197	229	269	338	$4·5·4·6·4·8_2$

Secondary building units: 1-4-1 or 4-[1,1]

Composite building units:
     d4r

Materials with this framework type:
     *ITQ-12[1,2]

ITQ-12   **Type Material Data**   **ITW**

Crystal chemical data: $[Si_{24}O_{48}]$-**ITW**
monoclinic, *Cm*

$$a = 10.3304\text{Å}, b = 15.010\text{Å}, c = 8.860\text{Å}, \beta = 105.34°\,^{(1)}$$

Framework density: $18.1 \text{ T}/1000\text{Å}^3$

Channels: [100] **8** 2.4 x 5.4* ↔ [001] **8** 3.9 x 4.2*

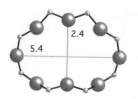 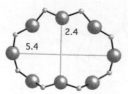 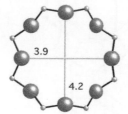 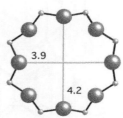

       8-ring viewed along [100]   *8-ring viewed along [001]*

References:
(1) Barrett, P.A., Boix, T., Puche, M., Olson, D.H., Jordan, E., Koller, H. and Camblor, M.A *Chem. Commun.*, 2114–2115 (2003)
(2) Yang, X.B., Camblor, M.A., Lee, Y., Liu, H.M. and Olson, D.H. *J. Am. Chem. Soc.*, **126**, 10403–10409 (2004)

IWR

Cmmm

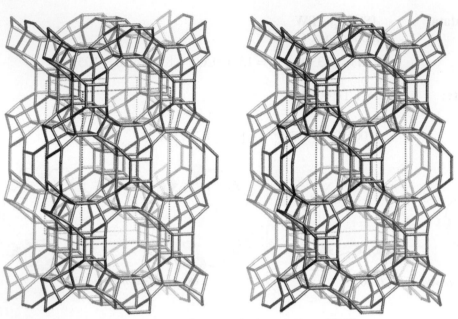

framework viewed along [001]

Idealized cell data: orthorhombic, *Cmmm*, $a = 21.2$Å, $b = 13.3$Å, $c = 12.7$Å

Coordination sequences and vertex symbols:

T_1 (16,1)	4	9	18	32	50	72	98	128	158	192	240	291	$4·6·4·6_2·4·10_4$
T_2 (16,1)	4	11	18	31	49	72	96	125	159	194	237	285	$4·6_2·5·6·5·10_2$
T_3 (16,1)	4	10	19	32	49	69	98	132	161	190	233	289	$4·4·5·6_2·5·12_7$
T_4 (8,..*m*)	4	11	20	27	43	73	99	117	147	198	246	281	$4·5_2·6·6_2·6·6_2$

Secondary building units: 1-5-1

Composite building units:

d4r	bre	lau	mel

Materials with this framework type:
   *ITQ-24[1]

ITQ-24   **Type Material Data**   **IWR**

Crystal chemical data: $[Al_{2.6}Ge_{5.1}Si_{48.3}O_{224}]$-**IWR**
orthorhombic, *Cmmm*, $a = 21.2549$Å, $b = 13.5210$Å, $c = 12.6095$Å [(1)]

Framework density: 15.5 T/1000Å^3

Channels: [001] **12** 5.8 x 6.8* ↔ [110] **10** 4.6 x 5.3* ↔ [010] **10** 4.6 x 5.3*
(The 10-ring shown below (viewed along [010]) limits the channel
dimensions of both 10-ring channels.)

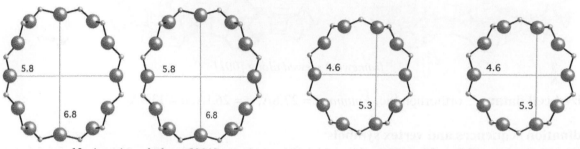

           12-ring viewed along [001]   *10-ring viewed along [010]*

References:
(1)   Castaneda, R., Corma, A., Fornes, V., Rey, F. and Rius, J. *J. Am. Chem. Soc.*, **125**, 7820–7821 (2003)

IWV

Fmmm

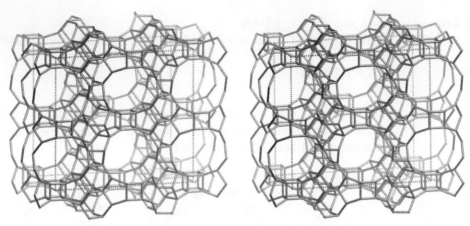

framework viewed along [001]

Idealized cell data: orthorhombic, *Fmmm*, a = 27.8Å, b = 26.1Å, c = 13.9Å

Coordination sequences and vertex symbols:

T_1 (32,1)	4	12	20	32	50	74	101	135	167	203	254	307	$5 \cdot 5 \cdot 5 \cdot 6_2 \cdot 5 \cdot 12$
T_2 (32,1)	4	9	18	32	52	78	105	130	164	213	264	310	$4 \cdot 5 \cdot 4 \cdot 5 \cdot 4 \cdot 12$
T_3 (32,1)	4	11	21	34	49	72	101	138	177	204	243	292	$4 \cdot 6 \cdot 5 \cdot 5 \cdot 5_2 \cdot 12$
T_4 (16,..m)	4	12	20	34	50	67	100	141	178	214	232	278	$5 \cdot 5_2 \cdot 5 \cdot 5_2 \cdot 14_2 \cdot *$
T_5 (16,..m)	4	12	20	28	49	69	100	136	166	201	245	292	$5 \cdot 5 \cdot 5 \cdot 5 \cdot 5 \cdot 6_2$
T_6 (16,.m.)	4	12	22	32	45	69	101	137	167	199	244	303	$5_2 \cdot 5 \cdot 6 \cdot 5 \cdot 6$
T_7 (8,2mm)	4	12	24	32	40	66	108	136	168	196	240	298	$5 \cdot 5 \cdot 5 \cdot 5 \cdot 14_6 \cdot *$

Secondary building units: see *Compendium*

Composite building units:

    d4r   *cas*   *non*   *ton*

Materials with this framework type:
    *ITQ-27[1]

ITQ-27   **Type Material Data**   **IWV**

Crystal chemical data: [Al$_5$Si$_{147}$O$_{304}$]-**IWV**
                              orthorhombic, *Fmmm*, $a = 27.7508$Å, $b = 25.2969$Å, $c = 13.7923$Å [1]

Framework density: 15.7 T/1000Å^3

Channels: { [001] **12** 6.2 x 6.9 ↔ [011] **12** 6.2 x 6.9 }**
                              (The channel dimensions are limited by the same 12-ring in both
                              directions.)

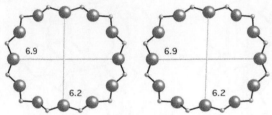

12-ring viewed along [001]

References:
(1) Dorset, D.L., Kennedy, G.J., Strohmaier, K.G., Diaz-Cabanas, M.J., Rey, F. and Corma, A. *J. Am. Chem. Soc.*, **128**, 8862–8867 (2006)

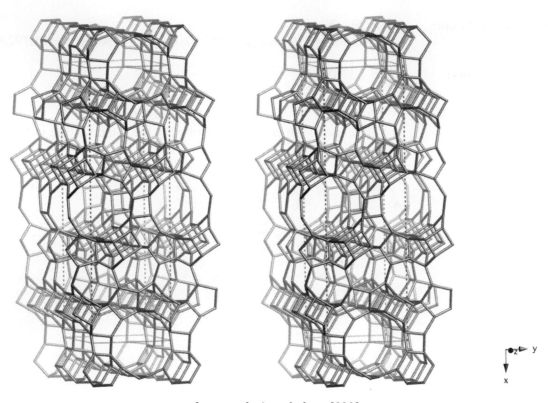

framework viewed along [001]

Idealized cell data: orthorhombic, *Pbam*, $a = 41.7$Å, $b = 12.7$Å, $c = 12.7$Å

Coordination sequences and vertex symbols:
   see Appendix A for a list of the coordination sequences and vertex symbols for the 16 T-atoms

Secondary building units: 1-5-1

Composite building units:

| *d4r* | *mor* | *bre* | *lau* | *stf* | *mel* |

Materials with this framework type:
   *ITQ-22[1]

Crystal chemical data: $[Ge_{22.2}Si_{89.8}O_{224}]$-**IWW**
orthorhombic, *Pbam*, $a = 42.1326$Å, $b = 12.9885$Å, $c = 12.6814$Å [1]

Framework density: 16.1 T/1000Å^3

Channels: [001] **12** 6.0 x 6.7* ↔ ⊥ [001] **10** 4.9 x 4.9** ↔ [001] **8** 3.3 x 4.6*

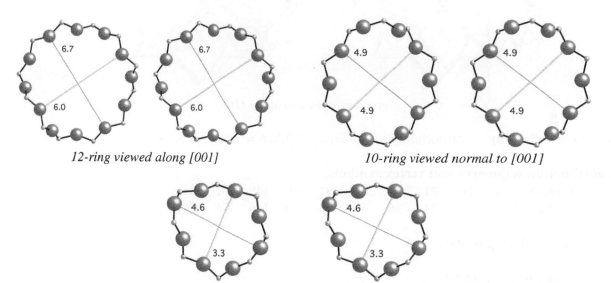

12-ring viewed along [001]   *10-ring viewed normal to [001]*

8-ring viewed allong [001]

References:
(1)   Corma, A., Rey, F., Valencia, S., Jorda, J.L. and Rius, J. *Nature Materials*, **2**, 493–497 (2003)

JBW Framework Type Data *Pmma*

framework viewed along [100]

Idealized cell data: orthorhombic, *Pmma*, *a* = 5.3Å, *b* = 7.5Å, *c* = 8.2Å

Coordination sequences and vertex symbols:

T_1 (4,*m*)	4	10	21	39	61	81	107	148	192	228	$4{\cdot}6_2{\cdot}4{\cdot}6_2{\cdot}6{\cdot}8_2$
T_2 (2,*mm*2)	4	12	24	36	56	86	118	146	176	228	$6{\cdot}6{\cdot}6{\cdot}6{\cdot}6_2{\cdot}6_2$

Secondary building units: 6

Composite building units:

     dzc   *jbw*   *abw*

    double zigzag
       chain

Materials with this framework type:
  *Na-J (Barrer and White)[1]
  |Na_2 Rb H_2O|[Al_3Ge_3O_12]-**JBW**[2]
  |Na_3 (H_2O)_2|[Al_3Ge_3O_12]-**JBW**[3]

                          Nepheline hydrate[4]
                          |Na-|[Al-Si-O]-**JBW**[5]

Na-J (Barrer and White)   Type Material Data   **JBW**

Crystal chemical data: $|Na_3 (H_2O)_{1.5}| [Al_3Si_3O_{12}]$-**JBW**
orthorhombic, $Pna2_1$, $a = 16.426$Å, $b = 15.014$Å, $c = 5.224$Å [1]
(Relationship to unit cell of Framework Type: $a' = 2c$, $b' = 2b$, $c' = a$)

Framework density: 18.6 T/1000Å^3

Channels: [001] **8** 3.7 x 4.8*

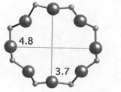 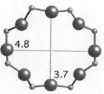

8-ring along [001]

References:
(1) Hansen, S. and Fälth, L. *Zeolites*, **2**, 162–166 (1982)
(2) Healey, A.M., Henry, P.F., Johnson, G.M., Weller, M.T., Webster, M. and Genge, A.J. *Microporous Mesoporous Mat.*, **37**, 165–174 (2000)
(3) Tripathi, A. and Parise, J.B. *Microporous Mesoporous Mat.*, **52**, 65–78 (2002)
(4) Rheinhardt, A., Hellner, E. and Ahsbahs, H. *Fortsch. Mineral.*, **60**, 175–176 (1982)
(5) Ragimov, K.G., Chiragove, M.I., Mustafaev, N.M. and Mamedov, Kh.S. *Sov. Phys. Dokl.*, **23**, 697–698 (1978)

KFI

Framework Type Data

Im̄3m

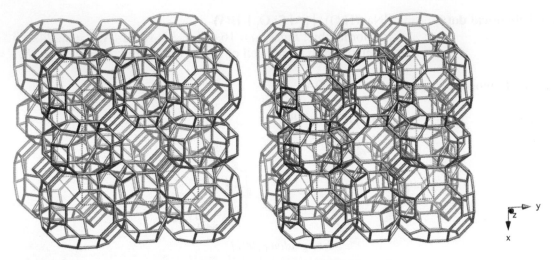

framework viewed along [001]

Idealized cell data: cubic, $Im\overline{3}m$, $a = 18.6$Å

Coordination sequences and vertex symbols:

    T_1 (96,1)   4   9   17   29   45   64   86   112   141   173   4·4·4·8·6·8

Secondary building units: 6-6 or 6-2 or 8 or 6 or 4-2 or 4

Composite building units:

        d6r   *pau*   *lta*

  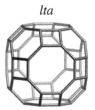

Materials with this framework type:

*ZK-5[1]

(Cs,K)-ZK-5[2,3]

[Zn-Ga-As-O]-**KFI**[4]

|18-crown-6|[Al-Si-O]-**KFI**[5]

P[6]

Q[6]

Crystal chemical data: $|Na_{30}(H_2O)_{98}|[Al_{30}Si_{66}O_{192}]$-**KFI**
cubic, $Im\overline{3}m$, $a = 18.75\text{Å}$ [1]

Framework density: $14.6\ T/1000\text{Å}^3$

Channels: <100> **8** 3.9 x 3.9*** | <100> **8** 3.9 x 3.9***

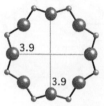   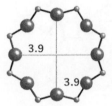

8-ring viewed along <100>

References:
(1) Meier, W.M. and Kokotailo, G.T. *Z. Kristallogr.*, **121**, 211–219 (1965)
(2) Robson, H.E. *U.S. Patent 3,720,753* (1973)
(3) Parise, J.B., Shannon, R.D., Prince, E. and Cox, D.E. *Z. Kristallogr.*, **165**, 175–190 (1983)
(4) Feng, P., Zhang, T. and Bu, X. *J. Am. Chem. Soc.*, **123**, 8608–8609 (2001)
(5) Chatelain, T., Patarin, J., Farre, R., Petigny, O. and Schulz, P. *Zeolites*, **17**, 328–333 (1996)
(6) Barrer, R.M. and Robinson, D. *Z. Kristallogr.*, **135**, 374–390 (1972)

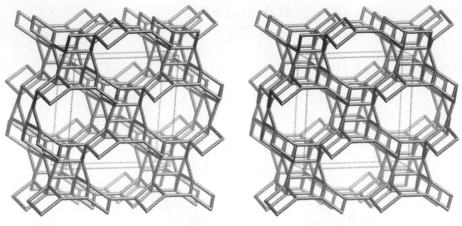

framework viewed along [001]

Idealized cell data: monoclinic, $C2/m$, $a = 14.6$Å, $b = 12.9$Å, $c = 7.6$Å, $\beta = 111.2°$

Coordination sequences and vertex symbols:

T_1 (8,1)	4	10	20	33	51	74	99	128	161	199	$4\cdot4\cdot6\cdot6_2\cdot6\cdot10_4$
T_2 (8,1)	4	10	19	32	52	74	99	126	162	203	$4\cdot4\cdot6\cdot6_2\cdot6\cdot10_2$
T_3 (8,1)	4	10	19	33	53	74	96	127	166	201	$4\cdot4\cdot6\cdot6_3\cdot6\cdot6_3$

Secondary building units: 6 or 1-4-1

Composite building units:

        bog   *lau*

Materials with this framework type:

*Laumontite[1-4]

[Co-Ga-P-O]-**LAU**[5,6]

[Fe-Ga-P-O]-**LAU**[6]

[Mn-Ga-P-O]-**LAU**[6]

[Zn-Al-As-O]-**LAU**[7]

[Zn-Ga-P-O]-**LAU**[8]

Leonhardite[9,10]

Na,K-rich laumontite[11]

Primary leonhardite[12]

Synthetic laumontite[13]

Laumontite   Type Material Data   **LAU**

Crystal chemical data: $|Ca_4 (H_2O)_{16}|$ $[Al_8Si_{16}O_{48}]$-**LAU**

monoclinic, *Am*, $a = 7.549$Å, $b = 14.740$Å, $c = 13.072$Å, $\gamma = 111.9°$ [3]
(Relationship to unit cell of Framework Type:

$a' = c, b' = a, c' = b, \gamma' = \beta$)

Framework density: 17.8 T/1000Å^3

Channels: [100] **10** 4.0 x 5.3* (contracts upon dehydration)

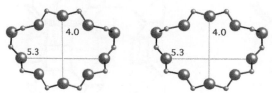

10-ring viewed along [100]

References:
(1)   Bartl, H. and Fischer, K. *N. Jb. Miner. Mh.*, 33–42 (1967)
(2)   Amirov, S.T., Ilyukhin, V.V. and Belov, N.V. *Dokl. Akad. Nauk SSSR*, **174**, 667– (1967)
(3)   Schramm, V. and Fischer *Adv. Chem. Ser.*, **101**, 259–265 (1971)
(4)   Artioli, G. and Ståhl, K. *Zeolites*, **17**, 249–255 (1993)
(5)   Chippindale, A.M. and Walton, R.I. *Chem. Commun.*, 2453–2454 (1994)
(6)   Bond, A.D., Chippindale, A., Cowley, A.R., Readman, J.E. and Powell, A.V. *Zeolites*, **19**, 326–333 (1997)
(7)   Feng, P., Zhang, T. and Bu, X. *J. Am. Chem. Soc.*, **123**, 8608–8609 (2001)
(8)   Cowley, A.R., Jones, R.H., Teat, S.J. and Chippindale, A.M. *Microporous Mesoporous Mat.*, **51**, 51–64 (2002)
(9)   Lapham, D.L. *Am. Mineral.*, **48**, 683–689 (1963)
(10) Artioli, G., Smith, J.V. and Kvick, Å *Zeolites*, **9**, 377–391 (1989)
(11) Stolz, J. and Armbruster, T. *N. Jb. Miner. Mh.*, 131–144 (1997)
(12) Baur, W.H., Joswig, W., Fursenko, B.A. AND Belitsky, I.A. *Eur. J. Mineral.*, **9**, 1173–1182 (1997)
(13) Ghobarkar, H. and Schaef, O. *Microporous Mesoporous Mat.*, **23**, 55–60 (1998)

LEV

Framework Type Data

$R\bar{3}m$

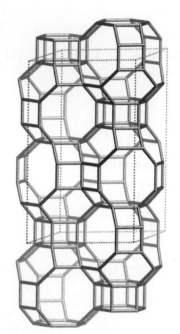 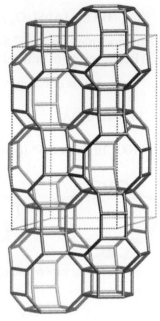 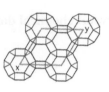

framework viewed normal to [001] (upper right: projection down [001])

Idealized cell data: trigonal, $R\bar{3}m$, $a = 13.2$Å, $c = 22.6$Å

Coordination sequences and vertex symbols:

T_1 (36,1)	4	9	17	30	49	71	92	114	143	183	4·4·4·6·6·8
T_2 (18,2)	4	10	20	32	46	64	90	124	156	184	4·4·6·6·8·8

Secondary building units: 6

Framework description: AABCCABBC sequence of 6-rings

Composite building units:
d6r

Materials with this framework type:

*Levyne[1,2]	LZ-132[5]	SAPO-35[8]
AlPO-35[3]	NU-3[6]	ZK-20[9]
CoDAF-4[4]	RUB-1 ([B-Si-O]-**LEV**)[7]	ZnAPO-35[10]

Levyne   **Type Material Data**   **LEV**

Crystal chemical data: $|Ca_9(H_2O)_{50}|[Al_{18}Si_{36}O_{108}]$-**LEV**
                              trigonal, $R\bar{3}m$, $a = 13.338$Å, $c = 23.014$Å [2]

Framework density: 15.2 T/1000Å^3

Channels: $\perp$ [001] **8** 3.6 x 4.8**

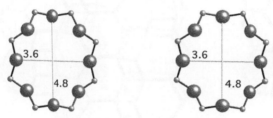

8-ring viewed normal to [001]

References:
(1) Barrer, R.M. and Kerr, I.S. *Trans. Faraday Soc.*, **55**, 1915–1923 (1959)
(2) Merlino, S., Galli, E. and Alberti, A. *Tschermaks Min. Petr. Mitt.*, **22**, 117–129 (1975)
(3) Zhu, G.S., Xiao, F.S., Qiu, S.L., Hun, P.C., Xu, R.R., Ma, S.J. and Terasaki, O.` *Microporous Materials*, **11**, 269–273 (1997)
(4) Barrett, P.A. and Jones, R.H. *Phys. Chem. Chem. Phys.*, **2**, 407–412 (2000)
(5) Tvaruzkova, Z., Tupa, M., Kiru, P., Nastro, A., Giordano, G. and Trifiro, F. *Int. Zeolite Sym., Wurzburg, Extended Abstracts* (1988)
(6) McCusker, L.B. *Mater. Sci. Forum*, **133-136**, 423–433 (1993)
(7) Grünewald-Luke, A., Marler, B., Hochgrafe, M. and Gies, H. *J. Mater. Chem.*, **9**, 2529–2536 (1999)
(8) Lok, B.M., Messina, C.A., Patton, R.L., Gajek, R.T., Cannan, T.R. and Flanigen, E.M. *J. Am. Chem. Soc.*, **106**, 6092–6093 (1984)
(9) Kerr, G.T. *U.S. Patent 3,459,676* (1969)
(10) Christensen, A.N. and Hazell, R.G. *Acta Chemica Scand.*, **53**, 403–409 (1999)

LIO

Framework Type Data

$P\bar{6}m2$

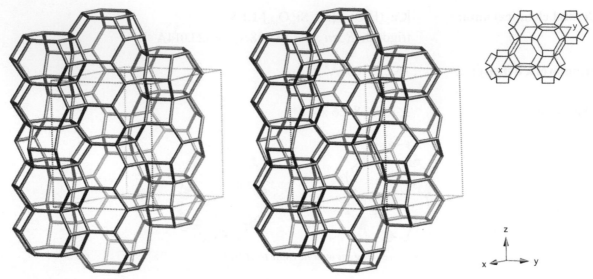

framework viewed normal to [001] (upper right: projection down [001])

Idealized cell data: hexagonal, $P\bar{6}m2$, $a = 12.3$Å, $c = 15.6$Å

Coordination sequences and vertex symbols:

T_1 (12,1)	4	10	20	34	53	76	103	135	170	209	4·6·4·6·6·6
T_2 (12,1)	4	10	20	34	54	78	104	134	168	210	4·4·6·6·6·6
T_3 (6,m)	4	10	20	34	52	74	102	136	172	208	4·6·4·6·6·6
T_4 (6,m)	4	10	20	34	54	78	104	134	168	210	4·6·4·6·6·6

Secondary building units: 6 or 4

Framework description: ABABAC sequence of 6-rings

Composite building units:

can   *los*   *lio*

   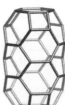

Materials with this framework type:
   *Liottite[1,2]

Liottite   Type Material Data   **LIO**

Crystal chemical data: $|Ca_8(K,Na)_{16}(SO_4)_5Cl_4|\ [Al_{18}Si_{18}O_{72}]$-**LIO**
hexagonal, $P\bar{6}$, $a = 12.870$Å, $c = 16.096$Å [2]

Framework density: 15.6 T/1000Å^3

Channels: apertures formed by 6-rings only

References:
(1) Merlino, S. and Orlandi, P. *Am. Mineral.*, **62**, 321–326 (1977)
(2) Ballirano, P., Merlino, S. and Bonaccorsi, E. *Can. Mineral.*, **34**, 1021–1030 (1996)

-LIT

Framework Type Data

Pnma

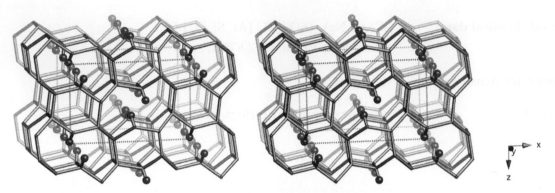

framework viewed along [010]

Idealized cell data: orthorhombic, *Pnma*, $a = 14.8$Å, $b = 8.6$Å, $c = 9.8$Å

Coordination sequences and vertex symbols:

T_1 (8,1)	3	7	16	31	50	68	93	128	161	196	237	283	$4 \cdot 6 \cdot 6$
T_2 (8,1)	4	9	19	34	50	72	100	128	160	197	239	287	$4 \cdot 8 \cdot 4 \cdot 8_2 \cdot 6 \cdot 6$
T_3 (8,1)	4	10	19	31	51	76	98	124	161	201	239	284	$4 \cdot 6 \cdot 6 \cdot 8_3 \cdot 8 \cdot 8$

Secondary building units: 6 or 4-[1,1] or 4-2

Materials with this framework type:
  *Lithosite[1]

Lithosite

Type Material Data

-LIT

Crystal chemical data: |K$_{12}$| [Al$_8$Si$_{16}$O$_{48}$(OH)$_4$]-**LIT**
monoclinic, $P2_1$

$a = 15.197$Å, $b = 10.233$Å, $c = 8.435$Å, $\beta = 90.31°$ [1]
(Relationship to unit cell of Framework Type: $a' = a$, $b' = c$, $c' = b$)

Framework density: 18.3 T/1000Å^3

Channels:

References:
(1)   Pudovkina, Z.V., Solov'eva, L.P. and Pyatenko, Yu.A. *Sov. Phys. Dokl.*, **31**, 941–942 (1986)

LOS

Framework Type Data

$P6_3/mmc$

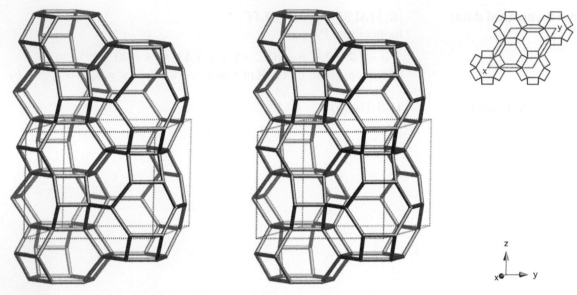

framework viewed normal to [001] (upper right: projection down [001])

Idealized cell data: hexagonal, $P6_3/mmc$, $a = 12.6$Å, $c = 10.3$Å

Coordination sequences and vertex symbols:

T₁ (12,*m*)	4	10	20	34	52	74	102	136	172	210	4·6·4·6·6·6
T₂ (12,2)	4	10	20	34	54	78	104	134	168	210	4·4·6·6·6·6

Secondary building units: 6-2 or 6 or 4

Framework description: ABAC sequence of 6-rings

Composite building units:

can

los

Materials with this framework type:
*Losod[1,2]
[Al-Ge-O]-**LOS**[3]
|Li-|[Be-P-O]-**LOS**[4]
Bystrite[5]

Losod   **Type Material Data**   **LOS**

Crystal chemical data: $|Na_{12} (H_2O)_{18}|$ $[Al_{12}Si_{12}O_{48}]$-**LOS**
hexagonal, $P6_3mc$, $a = 12.906$Å, $c = 10.541$Å $^{(2)}$

Framework density: 15.8 T/1000Å^3

Channels: apertures formed by 6-rings only

References:
(1) Sieber, W. and Meier, W.M. *Helv. Chim. Acta*, **57**, 1533–1549 (1974)
(2) Schicker, P. *Ph.D. Thesis, ETH, Zürich, Switzerland* (1988)
(3) Sokolov, Yu.A., Maksimov, B.A., Ilyukhin, V.V. and Belov, N.V. *Sov. Phys. Dokl.*, **23**, 789–791 (1978)
(4) Harrison, W.T.A., Gier, T.E. and Stucky, G.D. *Zeolites*, **13**, 242–248 (1993)
(5) Pobedimskaya, E.A., Terent'eva, L.F., Sapozhnikov, A.N., Kashaev, A.A. and Dorokhova, G.I. *Sov. Phys. Dokl.*, **36**, 553–555 (1991)

LOV

Framework Type Data

$P4_2/mmc$

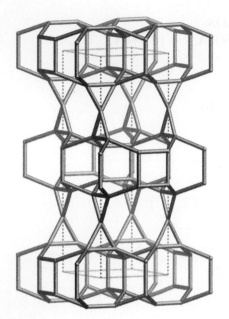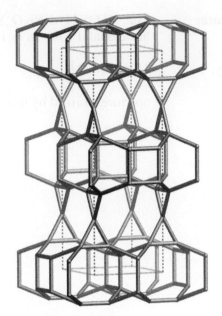

framework viewed normal [001]

Idealized cell data: tetragonal, $P4_2/mmc$, $a = 7.2$Å, $c = 20.9$Å

Coordination sequences and vertex symbols:

T_1 (8,m)	4	10	21	37	58	87	111	138	187	232	$4·4·6_2·8·6_2·8$
T_2 (8,m)	4	9	19	39	55	79	113	149	177	229	$3·4·8_3·9_4·8_3·9_4$
T_3 (2,$\overline{4}2m$)	4	8	20	40	54	76	116	144	200	210	$3·3·9_4·9_4·9_4·9_4$

Secondary building units: see *Compendium*

Composite building units:

lov   *vsv*

Materials with this framework type:
*Lovdarite[1,2]
Synthetic lovdarite[3]

Lovdarite   Type Material Data   **LOV**

Crystal chemical data: $|K_4Na_{12}(H_2O)_{18}|\,[Be_8Si_{28}O_{72}]$-**LOV**
orthorhombic, $Pma2$, $a = 39.576$Å, $b = 6.931$Å, $c = 7.153$Å $^{(2)}$
(Relationship to unit cell of Framework Type: $a' = 2c$, $b' = c' = a$)

Framework density: 18.3 T/1000Å^3

Channels: [010] **9** 3.2 x 4.5* ↔ [001] **9** 3.0 x 4.2* ↔ [100] **8** 3.6 x 3.7*

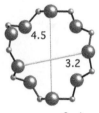
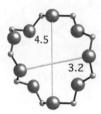

9-ring viewed along [010]

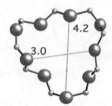
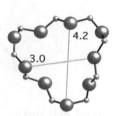

9-ring viewed along [001]

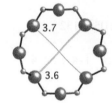
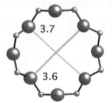

8-ring viewed along [100]

References:

(1) Merlino, S. *Acta Crystallogr. (Suppl.)*, **A37**, C189 (1981)
(2) Merlino, S. *Eur. J. Mineral.*, **2**, 809–817 (1990)
(3) Ueda, S., Koizumi, M., Baerlocher, Ch., McCusker, L.B. and Meier, W.M. *Preprints of Poster Papers, 7th Int. Zeolite Conf.*, pp. 23–24 (1986)

LTA

Framework Type Data

Pm$\bar{3}$m

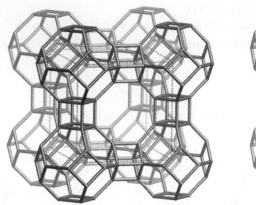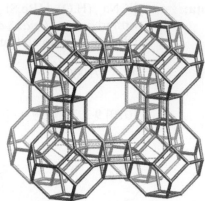

framework viewed along [001]

Idealized cell data: cubic, $Pm\bar{3}m$, $a = 11.9$Å

Coordination sequences and vertex symbols:

| T$_1$ (24,*m*) | 4 | 9 | 17 | 28 | 42 | 60 | 81 | 105 | 132 | 162 |

4·6·4·6·4·8

Secondary building units: 8 or 4-4 or 6-2 or 6 or 1-4-1 or 4

Composite building units:

d4r   *sod*   *lta*

Materials with this framework type:

*Linde Type A (zeolite A)[1,2]

[Al-Ge-O]-**LTA**[3]

[Ga-P-O]-**LTA**[4]

Alpha[5]

Dehyd. Linde Type A (dehyd. zeolite A)[6]

ITQ-29[7]

LZ-215[8]

N-A[9]

SAPO-42[10]

ZK-21[11]

ZK-22[11]

ZK-4[12]

Linde Type A   Type Material Data   **LTA**

Crystal chemical data: $|Na_{12} (H_2O)_{27}|_8 [Al_{12}Si_{12} O_{48}]_8$-**LTA**
cubic, $Fm\overline{3}c$, $a = 24.61Å^{(2)}$
(Relationship to unit cell of Framework Type: $a' = 2a$)

Framework density: 12.9 T/1000Å^3

Channels: <100> **8** 4.1 x 4.1***

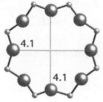 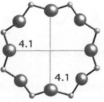

8-ring viewed along <100>

References:
(1) Reed, T.B. and Breck, D.W. *J. Am. Chem. Soc.*, **78**, 5972–5977 (1956)
(2) Gramlich, V. and Meier, W.M. *Z. Kristallogr.*, **133**, 134–149 (1971)
(3) Barrer, R.M., Baynham, J.W., Bultitude, F.W. and Meier, W.M. *J. Chem. Soc.*, 195–208 (1959)
(4) Simmen, A., Patarin, J. and Baerlocher, Ch. *Proc. 9th Int. Zeolite Conf.*, pp. 433–440 (1993)
(5) Wadlinger, R.L., Rosinski, E.J. and Plank, C.J. *U.S. Patent 3,375,205* (1968)
(6) Pluth, J.J. and Smith, J.V. *J. Am. Chem. Soc.*, **102**, 4704–4708 (1980)
(7) Corma, A., Rey, F., Rius, J., Sabater, M.J. and Valencia, S. *Nature*, **431**, 287–290 (2004)
(8) Breck, D.W. and Skeels, G.W. *U.S. Patent 4,503,023* (1985)
(9) Barrer, R.M. and Denny, P.J. *J. Chem. Soc.*, 971–982 (1961)
(10) Lok, B.M., Messina, C.A., Patton, R.L., Gajek, R.T., Cannan, T.R. and Flanigen, E.M. *J. Am. Chem. Soc.*, **106**, 6092–6093 (1984)
(11) Kuehl, G.H. *Inorg. Chem.*, **10**, 2488–2495 (1971)
(12) Kerr, G.T. *Inorg. Chem.*, **5**, 1537–1539 (1966)

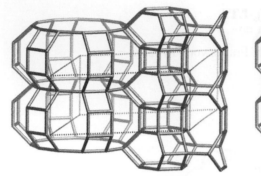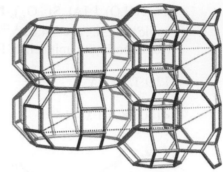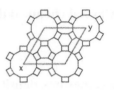

framework viewed normal to [001] (upper right: projection down [001])

Idealized cell data: hexagonal, *P6/mmm*, $a = 18.1$Å, $c = 7.6$Å

Coordination sequences and vertex symbols:

T_1 (24,1)	4	9	17	29	46	69	98	131	162	187	$4 \cdot 4 \cdot 4 \cdot 6 \cdot 6 \cdot 8$
T_2 (12,*m*)	4	10	21	35	49	66	89	117	150	190	$4 \cdot 8_3 \cdot 4 \cdot 8_3 \cdot 6 \cdot 12$

Secondary building units: 6 or 4-2

Composite building units:

dsc	d6r	can	ltl
double sawtooth chain			

Materials with this framework type:

(K,Ba)-G,L[1]
*Linde Type L (zeolite L)[2]
Gallosilicate L[3,4]

LZ-212[5]
Perlialite[6,7]
[Al-P-O]-**LTL**[8]

Crystal chemical data: $|K_6Na_3(H_2O)_{21}|\,[Al_9Si_{27}O_{72}]$-**LTL**
hexagonal, $P6/mmm$, $a = 18.40$Å, $c = 7.52$Å [2]

Framework density: 16.3 T/1000Å^3

Channels: [001] **12** 7.1 x 7.1*

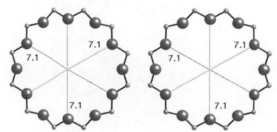

12-ring viewed along [001]

References:
(1) Baerlocher, Ch. and Barrer, R.M. *Z. Kristallogr.*, **136**, 245–254 (1972)
(2) Barrer, R.M. and Villiger, H. *Z. Kristallogr.*, **128**, 352–370 (1969)
(3) Wright, P.A., Thomas, J.M., Cheetham, A.K. and Nowak, A.K. *Nature*, **318**, 611–614 (1985)
(4) Newsam, J.M. *Mater. Res. Bull.*, **21**, 661–672 (1986)
(5) Breck, D.W. and Skeels, G.W. *U.S. Patent 4,503,023* (1985)
(6) Menshikov, Y.P. *Zap. Vses. Mineral. O-va*, **113**, 607–612 (1984)
(7) Artioli, G. and Kvick, Å. *Eur. J. Mineral.*, **2**, 749–759 (1990)
(8) Venkatathri, N. *Indian J Chem Sect A*, **41**, 2223–2230 (2002)

LTN

Framework Type Data

Fd$\bar{3}$m

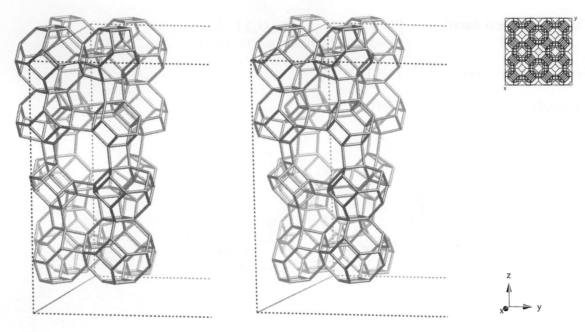

framework viewed normal to [001] (upper right: projection down [001])

Idealized cell data: cubic, Fd$\bar{3}$m (origin choice 2), a = 35.6Å

Coordination sequences and vertex symbols:

T$_1$ (192,1)	4	10	20	34	53	76	102	132	166	205	4·4·6·6·6·6
T$_2$ (192,1)	4	10	20	34	51	71	96	126	162	202	4·6·4·6·6·6
T$_3$ (192,1)	4	10	20	33	50	71	97	129	163	200	4·4·6·6·6·8
T$_4$ (192,1)	4	9	17	30	49	72	97	125	158	197	4·4·4·8·6·6

Secondary building units: 6 or 4-2

Composite building units:

d6r   can   sod   lta

Materials with this framework type:
   *Linde Type N[1]
   NaZ-21[2]

Linde Type N   Type Material Data   **LTN**

Crystal chemical data: $|Na_{384} (H_2O)_{518}| [Al_{384}Si_{384}O_{1536}]$-**LTN**
cubic, $Fd\overline{3}$, $a = 36.93Å$ [1]

Framework density: $15.2\ T/1000Å^3$

Channels: apertures formed by 6-rings only

References:
(1) Fälth, L. and Andersson, S. *Z. Kristallogr.*, **160**, 313–316 (1982)
(2) Shepelev, Yu.F., Smolin, Yu.I., Butikova, I.K. and Tarasov, V.I. *Dokl. Akad. Nauk SSSR*, **272**, 1133–1137 (1983)

MAR

Framework Type Data

$P6_3/mmc$

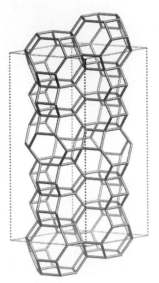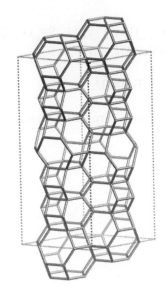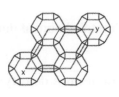

framework viewed normal to [001] (upper right: projection down [001])

Idealized cell data: hexagonal, $P6_3/mmc$, $a = 12.4$Å, $c = 30.7$Å

Coordination sequences and vertex symbols:

T_1 (24,1)	4	10	20	34	53	76	102	132	167	209	255	301	4·4·6·6·6·6
T_2 (24,1)	4	10	20	34	53	76	102	132	166	205	249	297	4·6·4·6·6·6
T_3 (12,m..)	4	10	20	34	54	78	104	134	168	208	252	298	4·6·4·6·6·6
T_4 (12,.2.)	4	10	20	34	52	74	102	136	172	208	248	298	4·4·6·6·6·6

Secondary building units: 8 or 4

Framework description: ABCBCBACBCBC sequence of 6-rings

Composite building units:

     can   *sod*   *lio*

Materials with this framework type:
    *Marinellite[1]

Marinellite **Type Material Data** **MAR**

Crystal chemical data: $|Na_{31}K_{11}Ca_6(H_2O)_{3.4}(SO_4)_8Cl_2|[Al_{36}Si_{36}O_{144}]$-**MAR**
trigonal, $P31c$, $a = 12.880$Å, $c = 31.761$Å $^{(1)}$

Framework density: 15.8 T/1000Å^3

Channels: apertures formed by 6-rings only

References:
(1) Bonaccorsi, E. and Orlandi, P. *Eur. J. Mineral.*, **15**, 1019–1027 (2003)

MAZ

Framework Type Data

P6₃/mmc

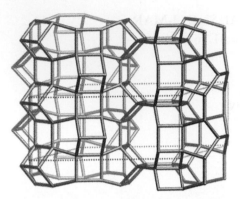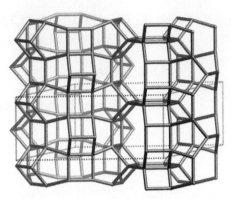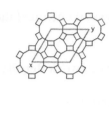

framework viewed normal to [001] (upper right: projection down [001])

Idealized cell data: hexagonal, $P6_3/mmc$, $a = 18.1$Å, $c = 7.6$Å

Coordination sequences and vertex symbols:

T₁ (24,1)	4	10	20	35	54	78	104	134	171	210		4·5·4·5·8·12
T₂ (12,*m*)	4	10	21	36	53	74	104	138	174	212		4·8₂·4·8₂·5·6

Secondary building units: 5-1 or 4-2

Composite building units:

        dsc   *gme*

double sawtooth chain

Materials with this framework type:

*Mazzite[1,2]

[Ga-Si-O]-**MAZ**[3]

LZ-202[4]

Mazzite-Na, Boron, CA[5]

Omega[6,7]

ZSM-4[8]

Mazzite   **Type Material Data**   **MAZ**

Crystal chemical data: $|(Na_2,K_2,Ca,Mg)_5\,(H_2O)_{28}|\,[Al_{10}Si_{26}O_{72}]$-**MAZ**
hexagonal, $P6_3/mmc$, $a = 18.392$Å, $c = 7.646$Å [(2)]

Framework density: 16.1 T/1000Å^3

Channels: [001] **12** 7.4 x 7.4* | [001] **8** 3.1 x 3.1***

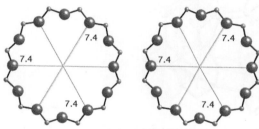
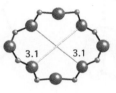
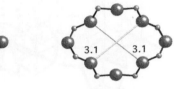

     12-ring viewed along [001]   *limiting 8-ring along [001]*

References:
(1) Galli, E. *Cryst. Struct. Comm.*, **3**, 339–344 (1974)
(2) Galli, E. *Rend. Soc. Ital. Mineral. Petrol.*, **31**, 599–612 (1975)
(3) Newsam, J.M., Jarman, R.H. and Jacobson, A.J. *Mater. Res. Bull.*, **20**, 125–136 (1985)
(4) Breck, D.W. and Skeels, G.W. *U.S. Patent 4,503,023* (1985)
(5) Arletti, R., Galli, E., Vezzalini, G. and Wise, W.S. *Am. Mineral.*, **90**, 1186–1191 (2005)
(6) Galli, E. *Cryst. Struct. Comm.*, **3**, 339–344 (1974)
(7) Martucci, A., Alberti, A., Guzman-Castillo, M.D., Di Renzo, F. and Fajula, F. *Microporous Mesoporous Mat.*, **63**, 33–42 (2003)
(8) Rubin, M.K., Plank, C.J. and Rosinski, E.J. *U.S. Patent 4,021,447* (1977)

MEI

Framework Type Data

$P6_3/m$

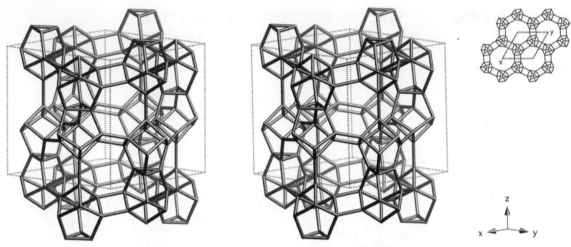

framework viewed normal to [001] (upper right: projection down [001])

Idealized cell data: hexagonal, $P6_3/m$, $a = 13.1$Å, $c = 15.6$Å

Coordination sequences and vertex symbols:

T_1 (12,1)	4	9	18	30	46	63	94	125	152	183	4·4·4·12·5·7
T_2 (12,1)	4	10	17	30	46	67	91	123	153	190	4·5·4·7·5·12
T_3 (6,m)	4	10	16	26	46	66	94	114	158	194	3·7·5·5·5·5
T_4 (4,3)	4	9	18	30	39	67	98	121	147	189	4·7·4·7·4·7

Secondary building units: see *Compendium*

Composite building units:
   mei

Materials with this framework type:
  *ZSM-18[1]
  ECR-40[2]

ZSM-18   **Type Material Data**   **MEI**

Crystal chemical data: $|Na_n(H_2O)_{28}|\ [Al_nSi_{34-n}O_{68}]$-**MEI**, n = 2.1 - 5.7
hexagonal, $P6_3/m$, $a = 13.175$Å, $c = 15.848$Å [1]

Framework density: 14.3 T/1000Å^3

Channels: [001] **12** 6.9 x 6.9* ↔ ⊥ [001] **7** 3.2 x 3.5**

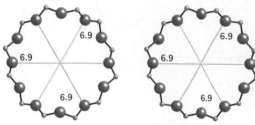

12-ring viewed along [001]   *7-ring viewed normal to [001]*

References:
(1) Lawton, S.L. and Rohrbaugh, W.J. *Science*, **247**, 1319–1321 (1990)
(2) Afeworki, M., Dorset, D.L., Kennedy, G.J and Strohmaier, K.G. *Stud. Surf. Sci. Catal.*, **154**, 1274–1281 (2004)

MEL

Framework Type Data

$I\bar{4}m2$

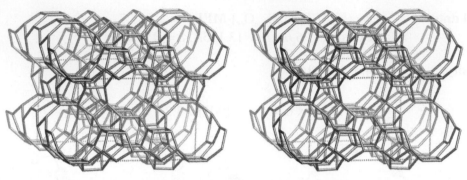

framework viewed along [100]

Idealized cell data: tetragonal, $I\bar{4}m2$, $a = 20.3$Å, $c = 13.5$Å

Coordination sequences and vertex symbols:

T_1 (16,1)	4	12	21	36	63	88	121	153	192	249	$5 \cdot 5 \cdot 5 \cdot 5 \cdot 5 \cdot 6$
T_2 (16,1)	4	11	23	38	61	93	121	153	198	246	$4 \cdot 5 \cdot 5 \cdot 6_2 \cdot 5 \cdot 10_2$
T_3 (16,1)	4	12	23	38	62	91	116	155	203	244	$5 \cdot 5 \cdot 5 \cdot 5_2 \cdot 5 \cdot 10_3$
T_4 (16,1)	4	12	23	38	59	87	122	158	198	243	$5 \cdot 6_2 \cdot 5 \cdot 10_2 \cdot 5_2 \cdot 6$
T_5 (16,1)	4	11	22	36	57	90	127	157	194	244	$4 \cdot 5 \cdot 5 \cdot 6 \cdot 5 \cdot 8_2$
T_6 (8,2)	4	11	23	38	57	88	126	158	191	236	$4 \cdot 5_2 \cdot 6_2 \cdot 6_2 \cdot 10 \cdot 10$
T_7 (8,2)	4	12	21	40	63	83	124	155	197	244	$5 \cdot 5 \cdot 5 \cdot 5 \cdot 5_2 \cdot 10_2$

Secondary building units: 5-1

Composite building units:

    mor   *mel*   *mfi*

Materials with this framework type:
*ZSM-11[1-3]
|DEOTA|[Si-B-O]-**MEL**[4]
Bor-D (**MFI/MEL**intergrowth)[5]
Boralite D[6]

SSZ-46[7,8]
Silicalite 2[9]
TS-2[10]

ZSM-11   **Type Material Data**   **MEL**

Crystal chemical data: |Na$_n$ (H$_2$O)$_{16}$| [Al$_n$Si$_{96-n}$O$_{192}$]-**MEL**, n < 16

tetragonal, $I\overline{4}m2$, $a = 20.12$Å, $c = 13.44$Å $^{(1)}$

Framework density: 17.6 T/1000Å^3

Channels: <100> **10** 5.3 x 5.4***

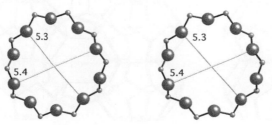

10-ring viewed along <100>

References:

(1) Kokotailo, G.T., Chu, P., Lawton, S.L. and Meier, W.M. *Nature*, **275**, 119–120 (1978)

(2) Fyfe, C.A., Gies, H., Kokotailo, G.T., Pasztor, C., Strobl, H. and Cox, D.E. *J. Am. Chem. Soc.*, **111**, 2470–2474 (1989)

(3) van Koningsveld, H., den Exter, M.J., Koegler, J.H., Laman, C.D., Njo, S.L. and Graafsma, H. *Proc. 12th Int. Zeolite Conf.*, **IV**, pp. 2419–2424 (1999)

(4) Piccione, P.M. and Davis, M.E. *Microporous Mesoporous Mat.*, **49**, 163–169 (2001)

(5) Perego, G. and Cesari, M. *J. Appl. Crystallogr.*, **17**, 403–410 (1984)

(6) Taramasso, M., Manara, G., Fattore, V. and Notari, B. *GB Patent 2,024,790* (1980)

(7) Terasaki, O., Ohsuna, T., Sakuma, H., Watanabe, D., Nakagawa, Y. and Medrud, R.C. *Chem. Mater.*, **8**, 463–468 (1996)

(8) Nakagawa, Y. and Dartt, C. *U.S. Patent 5,968,474* (1999)

(9) Bibby, D.M., Milestone, N.B. and Aldridge, L.P. *Nature*, **280**, 664–665 (1979)

(10) Reddy, J.S. and Kumar, R. *Zeolites*, **12**, 95–100 (1992)

MEP

$Pm\bar{3}n$

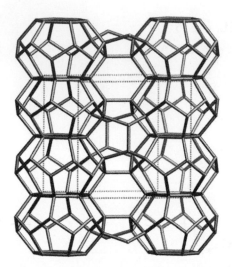 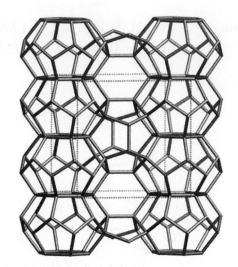

framework viewed along [001] (upper right: projection down [001])

Idealized cell data: cubic, $Pm\bar{3}n$, $a = 13.7$Å

Coordination sequences and vertex symbols:

T_1 (24,m)	4	12	25	42	69	100	129	176	229	277
T_2 (16,3)	4	12	24	42	67	95	133	177	219	277
T_3 (6, $\bar{4}2m$)	4	12	26	44	64	98	144	172	222	272

5·5·5·5·5·6
5·5·5·5·5·5
5·5·5·5·6·6

Secondary building units: see *Compendium*

Composite building units:
mtn

Materials with this framework type:
  *Melanophlogite[1]
  Synthetic melanophlogite[2]
  low melanophlogite[3]

Melanophlogite   **Type Material Data**

Crystal chemical data: | (CH$_4$,N$_2$,CO$_2$)$_x$| [Si$_{46}$O$_{92}$]-**MEP**
cubic, $Pm\overline{3}n$, $a =$ 13.436Å [1]

(Data refer to structure at 200°; tetragonal at 25°C)

Framework density: 19 T/1000Å^3

Channels: apertures formed by 6-rings only

References:
(1)   Gies, H. *Z. Kristallogr.*, **164**, 247–257 (1983)
(2)   Gies, H., Gerke, H. and Liebau, F. *N. Jb. Miner. Mh.*, 119–124 (1982)
(3)   Nakagawa, T., Kihara, K. and Harada, K. *Am. Mineral.*, **86**, 1506–1512 (2001)

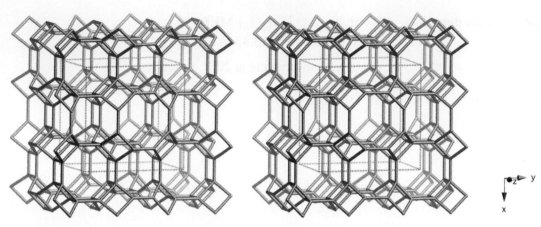

framework viewed along [001]

Idealized cell data: tetragonal, *I4/mmm*, *a* = 14.0Å, *c* = 10.0Å

Coordination sequences and vertex symbols:

T_1 (32,1) 4 9 18 32 49 69 93 121 153 189 $4 \cdot 4 \cdot 4 \cdot 8_2 \cdot 8 \cdot 8$

Secondary building units: 8-8 or 8 or 4

Composite building units:

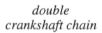
dcc

d8r

pau

*double
crankshaft chain*

Materials with this framework type:
  *Merlinoite[1,2]
  [Al-Co-P-O]-**MER**[3]
  [Ga-Al-Si-O]-**MER**[4]
  |Ba-|[Al-Si-O]-**MER**[5]
  |Ba-Cl-|[Al-Si-O]-**MER**[6]
  |K-|[Al-Si-O]-**MER**[7]

  |NH$_4$-|[Be-P-O]-**MER**[8]
  K-M[5,9]
  Linde W[5,10]
  Synthetic merlinoite[11]
  Zeolite W[12]

Merlinoite   **Type Material Data**   # MER

Crystal chemical data: $|K_5Ca_2 (H_2O)_{24}|$ $[Al_9Si_{23}O_{64}]$-**MER**
orthorhombic, *Immm*, $a = 14.116$Å, $b = 14.229$Å, $c = 9.946$Å [(2)]
(Relationship to unit cell of Framework Type: $a' = b' = a$, $c' = c$)

Framework density: 16 T/1000Å^3

Channels: [100] **8** 3.1 x 3.5* ↔ [010] **8** 2.7 x 3.6*

↔ [001] {**8** 3.4 x 5.1* + **8** 3.3 x 3.3*}

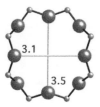
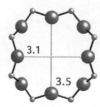
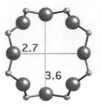
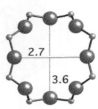

8-ring viewed along [100]   *8-ring viewed along [010]*

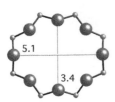
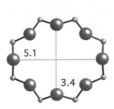
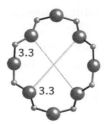
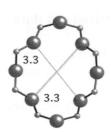

8-ring viewed along [001]   *2nd 8-ring viewed along [001]*

References:
(1) Passaglia, E., Pongiluppi, D. and Rinaldi, R. *N. Jb. Miner. Mh.*, 355–364 (1977)
(2) Galli, E., Gottardi, G. and Pongiluppi, D. *N. Jb. Miner. Mh.*, 1–9 (1979)
(3) Feng, P., Bu, X. and Stucky, G.D. *Nature*, **388**, 735–741 (1997)
(4) Kim, S.H., Kim, S.D., Kim, Y.C., Kim, C.S. and Hong, S.B. *Microporous Mesoporous Mat.*, **42**, 121–129 (2001)
(5) Gottardi, G. and Galli, E. *Natural Zeolites*, p. 157 (1985)
(6) Solov'eva, L.P., Borisov, S.V. and Bakakin, V.V. *Sov. Phys. Crystallogr.*, **16**, 1035–1038 (1972)
(7) Skofteland, B.M., Ellestad, O.H. and Lillerud, K.P. *Microporous Mesoporous Mat.*, **43**, 61–71 (2001)
(8) Bu, X., Gier, T.E. and Stucky, G.D. *Microporous Mesoporous Mat.*, **26**, 61–66 (1998)
(9) Barrer, R.M. and Baynham, J.W. *J. Chem. Soc.*, 2882–2891 (1956)
(10) Sherman, J.D. *ACS Sym. Ser.*, **40**, 30–42 (1977)
(11) Barrett, P.A., Valencia, S. and Camblor, M.A. *J. Mater. Chem.*, **8**, 2263–2268 (1998)
(12) Bieniok, A., Bornholdt, K., Brendel, U. and Baur, W.H. *J. Mater. Chem.*, **6**, 271–275 (1996)

MFI

Framework Type Data

Pnma

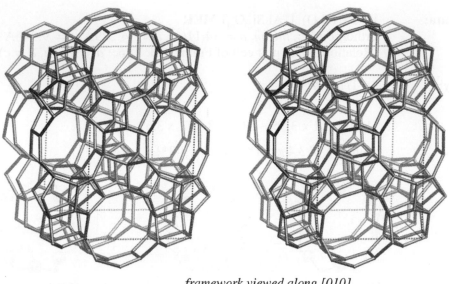

framework viewed along [010]

Idealized cell data: orthorhombic, *Pnma*, $a = 20.1$Å, $b = 19.7$Å, $c = 13.1$Å

Coordination sequences and vertex symbols:
see Appendix A for a list of the coordination sequences and vertex symbols for the 12 T-atoms

Secondary building units: 5-1

Composite building units:

mor	*cas*	*mel*	*mfi*

Materials with this framework type:

*ZSM-5[1-3]
[As-Si-O]-**MFI**[4]
[Fe-Si-O]-**MFI**[5]
[Ga-Si-O]-**MFI**[6]
AMS-1B[7]
AZ-1[8]
Bor-C[9]
Boralite C[10]
Encilite[11]

FZ-1[12]
LZ-105[13]
Monoclinic H-ZSM-5[14]
Mutinaite[15]
NU-4[16]
NU-5[17]
Silicalite[18]
TS-1[19]
TSZ[20]

TSZ-III[21]
TZ-01[22]
USC-4[23]
USI-108[24]
ZBH[25]
ZKQ-1B[26]
ZMQ-TB[27]
organic-free ZSM-5[28]

Crystal chemical data: $|Na_n (H_2O)_{16}| [Al_n Si_{96-n} O_{192}]$-**MFI** , n < 27

orthorhombic, *Pnma*, a = 20.07Å, b = 19.92Å, c = 13.42Å [(2)]

Framework density: 17.9 T/1000Å^3

Channels: {[100] **10** 5.1 x 5.5 ↔ [010] **10** 5.3 x 5.6}***

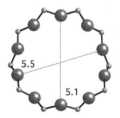 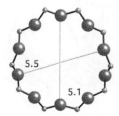 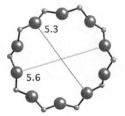 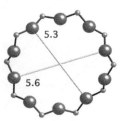

10-ring viewed along [100]   *10-ring viewed along [010]*

References:

(1) Kokotailo, G.T., Lawton, S.L., Olson, D.H. and Meier, W.M. *Nature*, **272**, 437–438 (1978)

(2) Olson, D.H., Kokotailo, G.T., Lawton, S.L. and Meier, W.M. *J. Phys. Chem.*, **85**, 2238–2243 (1981)

(3) van Koningsveld, H., van Bekkum, H. and Jansen, J.C. *Acta Crystallogr.*, **B43**, 127–132 (1987)

(4) Bhaumik, A. and Kumar, R. *Chem. Commun.*, 869–870 (1995)

(5) Patarin, J., Kessler, H. and Guth, J.L. *Zeolites*, **10**, 674–679 (1990)

(6) Awate, S.V., Joshi, P.N., Shiralkar, V.P. and Kotasthane, A.N. *J. Incl. Phenom.*, **13**, 207–218 (1992)

(7) Klotz, M.R. *U.S. Patent 4,269,813* (1981)

(8) Chono, M. and Ishida, H. *E. Patent B-113,116* (1984)

(9) Taramasso, M., Perego, G. and Notari, B. *Proc. 5th Int. Zeolite Conf.*, pp. 40–48 (1980)

(10) Taramasso, M., Manara, G., Fattore, V. and Notari, B. *GB Patent 2,024,790* (1980)

(11) Ratnasamy, P. and Borade, M.B. *E. Patent A-160,136* (1985)

(12) Suzuki, T., Hashimoto, S. and Nakano, R. *E. Patent B-31,255* (1981)

(13) Grose, R.W. and Flanigen, E.M. *U.S. Patent 4,257,885* (1981)

(14) van Koningsveld, H., Jansen, J.C. and van Bekkum, H. *Zeolites*, **10**, 235–242 (1990)

(15) Vezzalini, G., Quartieri, S., Galli, E., Alberti, A., Cruciani, G. and Kvick, A. *Zeolites*, **19**, 323–325 (1997)

(16) Whittam, T.V. *E. Patent B-65,401* (1986)

(17) Whittam, T.V. *E. Patent B-54,386* (1982)

(18) Flanigen, E.M., Bennett, J.M., Grose, R.W., Cohen, J.P., Patton, R.L., Kirchner, R.M. and Smith, J.V. *Nature*, **271**, 512–516 (1978)

(19) Taramasso, M., Perego, G. and Notari, B. *U.S. Patent 4,410,501* (1983)

(20) Ashibe, K., Kobayashi, W., Maejima, T., Sakurada, S. and Tagaya, N. *E. Patent A-101,232* (1984)

(21) Sakurada, S., Tagaya, N., Miura, T., Maeshima, T. and Hashimoto, T. *E. Patent A-170,751* (1986)

(22) Iwayama, K., Kamano, T., Tada, K. and Inoue, T. *E. Patent A-57,016* (1982)

(23) Young, D.A. *U.S. Patent 4,325,929* (1982)

(24) Hinnenkamp, J.A. and Walatka, V.V. *U.S. Patent 4,423,020* (1983)

(25) Holderich, W., Mross, W.D. and Schwartzmann, M. *E. Patent B-77,946* (1986)

(26) Kee Kwee, L.S.L. *E. Patent A-148,038* (1984)

(27) Kee Kwee, L.S.L. *E. Patent A-104,107* (1983)

(28) Kim, S.D., Noh, S.H., Seong, K.H. and Kim, W.J. *Microporous Mesoporous Mat.*, **72**, 185–192 (2004)

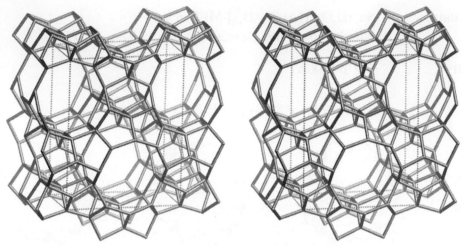

framework viewed along [010]

Idealized cell data: orthorhombic, *Imm2*, a = 7.5Å, b = 14.4Å, c = 19.0Å

Coordination sequences and vertex symbols:

T_1 (8,1)	4	12	23	39	62	93	127	165	210	260	$5 \cdot 5 \cdot 5 \cdot 5 \cdot 5 \cdot 8_2$
T_2 (8,1)	4	11	20	39	66	92	124	163	215	257	$4 \cdot 5_2 \cdot 5 \cdot 5 \cdot 8 \cdot 10$
T_3 (4,m)	4	12	24	42	64	90	131	168	206	259	$5 \cdot 5 \cdot 5 \cdot 5 \cdot 5 \cdot 10_2$
T_4 (4,m)	4	12	26	40	60	94	136	168	200	259	$5_2 \cdot 5_2 \cdot 6 \cdot 8 \cdot 6 \cdot 8$
T_5 (4,m)	4	12	19	35	64	96	123	155	207	272	$5 \cdot 5_2 \cdot 5 \cdot 5_2 \cdot 6 \cdot {}^*$
T_6 (4,m)	4	10	22	36	64	98	124	158	213	260	$4 \cdot 5 \cdot 4 \cdot 5 \cdot 8_2 \cdot 10$
T_7 (2,$mm2$)	4	12	22	40	62	92	138	160	196	262	$5 \cdot 5 \cdot 5 \cdot 5 \cdot 6_2 \cdot {}^*$
T_8 (2,$mm2$)	4	12	24	38	66	100	118	162	220	262	$5 \cdot 5 \cdot 5 \cdot 5 \cdot 5_2 \cdot 10_2$

Secondary building units: see *Compendium*

Composite building units:

mtt

fer

Materials with this framework type:
   *ZSM-57[1]
   COK-5 (intergrowth containing **MFS**)[2]

ZSM-57   **Type Material Data**   **MFS**

Crystal chemical data: $|H_{1.5}|$ $[Al_{1.5}Si_{34.5}O_{72}]$-**MFS**
orthorhombic, *Imm*2, $a = 7.451$Å, $b = 14.171$Å, $c = 18.767$Å [1]

Framework density: 18.2 T/1000Å^3

Channels: [100] **10** 5.1 x 5.4* ↔ [010] **8** 3.3 x 4.8*

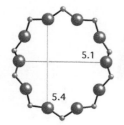 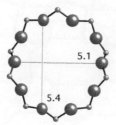 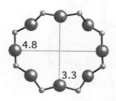 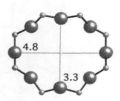

10-ring viewed along [100]   *8-ring viewed along [010]*

References:
(1) Schlenker, J.L., Higgins, J.B. and Valyocsik, E.W. *Zeolites*, **10**, 293–296 (1990)
(2) Kirschhock, C., Bons, A.J., Mertens, M., Ravishankar, R., Mortier, W., Jacobs, P. and Martens, J. *Chem. Mater.*, **17**, 5618–5624 (2005)

MON

*I*4₁/*amd*

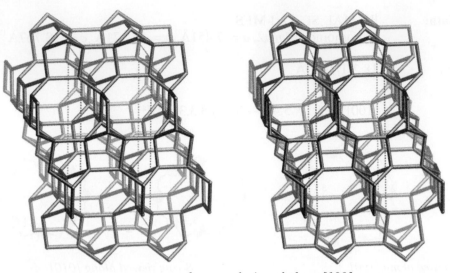

framework viewed along [100]

Idealized cell data: tetragonal, $I4_1/amd$ (origin choice 2), a = 7.1Å, c = 17.8Å

Coordination sequences and vertex symbols:

T₁ (16,*m*)   4   11   23   44   67   95   134   168   215   271 $4·5_2·5·8_2·5·8_2$

Secondary building units: 4

Materials with this framework type:
 *Montesommaite[1]
 [Al-Ge-O]-**MON**[2]5

Montesommaite   Type Material Data   **MON**

Crystal chemical data: $|(K,Na)_{4.5} (H_2O)_5| [Al_{4.5}Si_{11.5}O_{32}]$-**MON**
tetragonal, $I4_1/amd$, $a = 7.141$Å, $c = 17.307$Å [1]

Framework density: $18.1 \text{ T}/1000\text{Å}^3$

Channels: [100] **8** 3.2 x 4.4* ↔ [001] **8** 3.6 x 3.6*

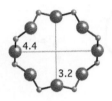
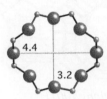

8-ring viewed along [100]

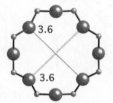
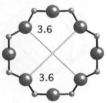

8-ring viewed along [001]

References:
(1) Rouse, R.C., Dunn, P.J., Grice, J.D., Schlenker, J.L. and Higgins, J.B. *Am. Mineral.*, **75**, 1415–1420 (1990)
(2) Tripathi, A. and Parise, J.B. *Microporous Mesoporous Mat.*, **52**, 65–78 (2002)

MOR

Framework Type Data

Cmcm

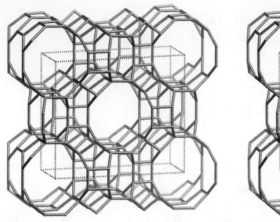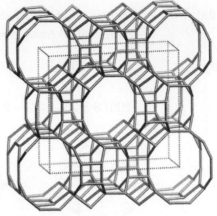

framework viewed along [001]

Idealized cell data: orthorhombic, *Cmcm*, *a* = 18.3Å, *b* = 20.5Å, *c* = 7.5Å

Coordination sequences and vertex symbols:

T_1 (16,1)	4	12	22	38	60	88	115	155	204	242		$5·5·5·5_2·8·12$
T_2 (16,1)	4	12	20	37	64	87	114	154	198	241		$5·5·5·5_2·5·8$
T_3 (8,*m*)	4	11	24	39	54	86	126	156	195	242		$4·5_2·5·8_2·5·8_2$
T_4 (8,*m*)	4	11	24	39	60	92	122	148	195	250		$4·5_2·5·8·5·8$

Secondary building units: 5-1

Composite building units:
mor

Materials with this framework type:

*Mordenite[1]

[Ga-Si-O]-**MOR**[2]

Ca-Q[3]

LZ-211[4]

Large port mordenite[5]

Maricopaite (interrupted framework)[6]

Mordenite, USA[7]

Na-D[8]

Mordenite Type Material Data **MOR**

Crystal chemical data: |Na$_8$ (H$_2$O)$_{24}$| [Al$_8$Si$_{40}$O$_{96}$]-**MOR**
orthorhombic, *Cmcm*, a = 18.1Å, b = 20.5Å, c = 7.5 Å [1]

Framework density: 17.2 T/1000Å^3

Channels: [001] **12** 6.5 x 7.0* ↔ [001] **8** 2.6 x 5.7***

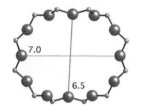 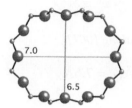 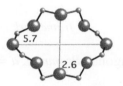 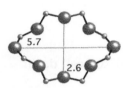

12-ring viewed along [001]   *limiting 8 ring along [001]*
between 12-ring channels

References:
(1) Meier, W.M. *Z. Kristallogr.*, **115**, 439–450 (1961)
(2) Eapen, M.J., Reddy, K.S.N., Joshi, P.N. and Shiralkar, V.P. *J. Incl. Phenom.*, **14**, 119–129 (1992)
(3) Koizumi, M. and Roy, R. *J. Geol.*, **68**, 41–53 (1960)
(4) Breck, D.W. and Skeels, G.W. *U.S. Patent 4,503,023* (1985)
(5) Sand, L.B. *Molecular Sieves*, pp. 71–77 (1968)
(6) Rouse, R.C. and Peacor, D.R. *Am. Mineral.*, **79**, 175–184 (1994)
(7) Gramlich, V. *Ph.D. Thesis, ETH, Zürich, Switzerland* (1971)
(8) Barrer, R.M. and White, E.A.D. *J. Chem. Soc.*, 1561–1571 (1952)

MOZ

Framework Type Data

P6/mmm

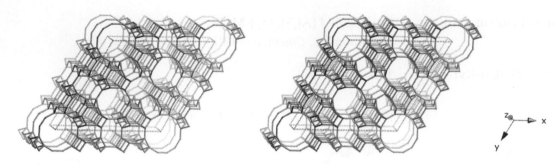

framework viewed along [001]

Idealized cell data: hexagonal, *P6/mmm*, $a = 31.2$Å, $c = 7.6$Å

Coordination sequences and vertex symbols:

T_1 (24,1)	4	9	17	29	46	70	100	130	155	179	215	271	$4 \cdot 4 \cdot 4 \cdot 6 \cdot 6 \cdot 8$
T_2 (24,1)	4	9	17	29	47	72	101	130	158	190	227	270	$4 \cdot 4 \cdot 4 \cdot 6 \cdot 6 \cdot 8$
T_3 (24,1)	4	9	17	31	53	78	100	122	151	190	237	287	$4 \cdot 4 \cdot 4 \cdot 6 \cdot 6 \cdot 8$
T_4 (12,*m*..)	4	10	21	35	49	66	90	121	157	195	233	274	$4 \cdot 8_3 \cdot 4 \cdot 8_3 \cdot 6 \cdot 12$
T_5 (12,*m*..)	4	10	21	34	46	64	92	127	169	211	245	272	$4 \cdot 8_3 \cdot 4 \cdot 8_3 \cdot 6 \cdot 8$
T_6 (12,*m*..)	4	10	21	35	51	71	94	124	163	204	243	284	$4 \cdot 8 \cdot 4 \cdot 8 \cdot 6 \cdot 8$

Secondary building units: 6 or 4-2

Composite building units:

dsc	*d6r*	*can*	*pau*	*ltl*
double sawtooth chain				

Materials with this framework type:
   *ZSM-10[1-3]

ZSM-10   **Type Material Data**   **MOZ**

Crystal chemical data: $|K_{24}(H_2O)_x|[Al_{24}Si_{84}O_{216}]$-**MOZ**
hexagonal, $P6/mmm$, $a = 31.575$ Å, $c = 7.525$ Å [3]

Framework density: 16.6 T/1000Å³

Channels: {[001] **12** 6.8 x 7.0 ↔ n[001] **8** 3.8 x 4.8}*** | [001] **12** 6.8 x 6.8*

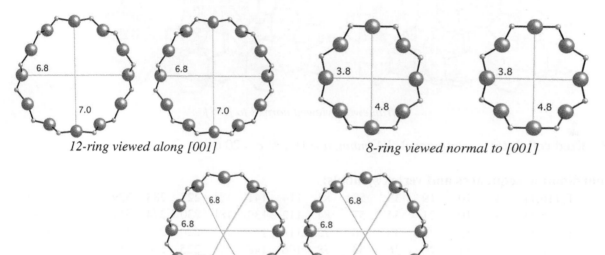

12-ring viewed along [001]   *8-ring viewed normal to [001]*

2nd 12-ring viewed along [001]

References:
(1) Higgins, J.B. and Schmitt, K.D. *Zeolites*, **16**, 236–244 (1996)
(2) Foster, M.D., Treacy, M.M.J., Higgins, J.B., Rivin, I., Balkovsky, E. and Randall, K.H. *J. Appl. Crystallogr.*, **38**, 1028–1030 (2005)
(3) Dorset, D.L. *Z. Kristallogr.*, **221**, 260–265 (2006)

MSE

Framework Type Data

P4₂/mnm

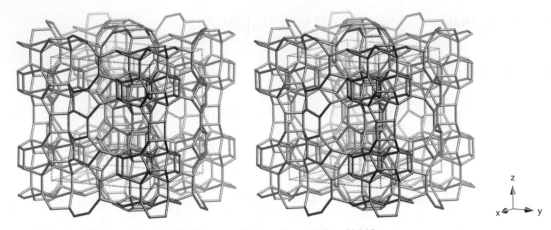

framework viewed normal to [001]

Idealized cell data: tetragonal, $P4_2/mnm$, $a = 18.2$Å, $c = 20.6$Å

Coordination sequences and vertex symbols:

T_1 (16,1)	4	10	19	33	53	83	114	141	174	227	283	329		4·5·4·6·5·6
T_2 (16,1)	4	10	21	34	57	84	112	139	181	230	274	317		4·5·4·10·5·6
T_3 (16,1)	4	12	22	33	50	75	114	152	183	220	267	324		5·6·5₂·6·6·10
T_4 (16,1)	4	11	20	36	54	78	106	148	193	225	254	317		4·5₂·5·5·5·10
T_5 (16,1)	4	12	18	35	58	81	110	143	177	230	271	328		5·5·5·6·5₂·10
T_6 (16,1)	4	10	20	34	56	83	109	140	181	226	282	337		4·5·4·10·5·6
T_7 (8,..*m*)	4	11	21	32	47	75	114	156	189	215	262	329		4·5·5·6·5·6
T_8 (8,..*m*)	4	11	20	31	55	85	115	141	179	223	275	331		4·5₂·5·6·5·6

Secondary building units: see *Compendium*

Composite building units:

    mor   *bea*   *mtw*

Materials with this framework type:
    *MCM-68[1]

MCM-68   **Type Material Data**   **MSE**

Crystal chemical data: $[Al_{11.4}Si_{100.6}O_{224}]$-**MSE**
                                      tetragonal, $P4_2/mnm$, $a = 18.286$Å, $c = 20.208$Å [1]

Framework density: 16.6 T/1000Å^3

Channels: {[001] **12** 6.4 x 6.8 ↔ [100] **10** 5.2 x 5.8 ↔ [110] **10** 5.2 x 5.2 }***

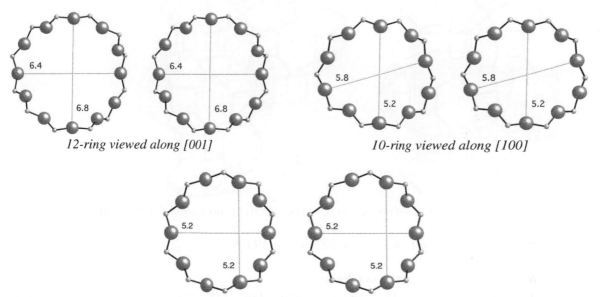

          12-ring viewed along [001]   *10-ring viewed along [100]*

10-ring viewed along [110]

References:
(1)   Dorset, D.L., Weston, S.C. and Dhingra, S.S. *J. Phys. Chem. B*, **110**, 2045–2050 (2006)

MSO

Framework Type Data

$R\bar{3}m$

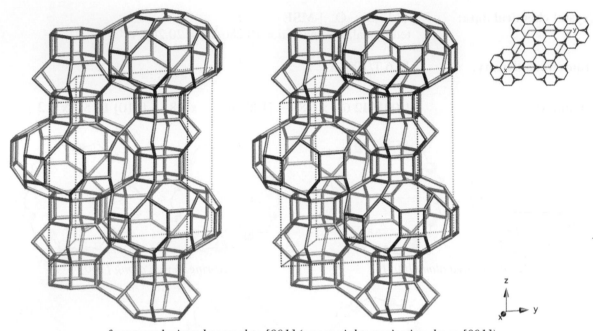

framework viewed normal to [001] (upper right: projection down [001])

Idealized cell data: trigonal, $R\bar{3}m$, $a = 17.2$Å, $c = 19.8$Å

Coordination sequences and vertex symbols:

T_1 (36,1)	4	10	21	37	55	75	101	136	175	211	$4 \cdot 6 \cdot 4 \cdot 6_2 \cdot 6 \cdot 6$
T_2 (36,1)	4	10	20	34	53	77	106	138	170	206	$4 \cdot 6_2 \cdot 4 \cdot 6_2 \cdot 6 \cdot 6$
T_3 (18,2)	4	12	21	32	51	80	110	132	164	212	$6 \cdot 6 \cdot 6 \cdot 6_2 \cdot 6_2 \cdot 6_2$

Secondary building units: 2-6-2 or 4-1

Composite building units:

d6r   lau   mso

Materials with this framework type:
    *MCM-61[1,2]
    Mu-13[3]

MCM-61 **Type Material Data** **MSO**

Crystal chemical data: $|K_{2.1} C_{12}H_{24}O_6| [Al_{2.1}Si_{27.9}O_{60}]$-**MSO**
$C_{12}H_{24}O_6$ = 18-crown-6

rhombohedral, $R\bar{3}m$, a = 11.841Å, α = 93.29°[(2)]
(hexagonal setting: a = 17.220Å, c = 19.296Å)

Framework density: 18.2 T/1000Å³

Channels: apertures formed by 6-rings only

References:
(1) Valyosik, E.W. *U.S. Patent 5,670,131* (1997)
(2) Shantz, D.F., Burton, A. and Lobo, R.F. *Microporous Mesoporous Mat.*, **31**, 61–73 (1999)
(3) Paillaud, J.-L., Caullet, P., Schreyeck, L. and Marler, B. *Microporous Mesoporous Mat.*, **42**, 177–189 (2001)

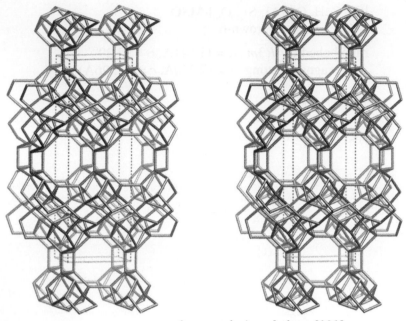

framework viewed along [001]

Idealized cell data: monoclinic, *C2/m*, *a* = 9.6Å, *b* = 30.4Å, *c* = 7.2Å, β = 90.5°

Coordination sequences and vertex symbols:

T_1 (8,1)	4	11	24	45	77	109	137	174	224	280	$4·5·5·6·5·8$
T_2 (8,1)	4	12	24	42	70	95	136	184	227	277	$5·5_2·5·6·5·8$
T_3 (8,1)	4	12	27	47	69	99	142	184	227	281	$5·6·5·7·5·8$
T_4 (8,1)	4	12	26	45	64	96	134	186	230	290	$5·6·5·6·5·6$
T_5 (8,1)	4	12	24	42	64	93	133	179	234	290	$5·6·5·6_2·5·7$
T_6 (4,2)	4	12	21	44	74	106	138	172	226	284	$5·5·5·5·5_2·6$

Secondary building units: see *Compendium*

Composite building units:

mfi

Materials with this framework type:
    *MCM-35[1]
    UTM-1[2]

227

MCM-35   **Type Material Data**   **MTF**

Crystal chemical data: [Si$_{44}$O$_{88}$]-**MTF**
                               monoclinic, C2/m

                               $a = 9.500$ Å, $b = 30.710$Å, $c = 7.313$Å, $\beta = 91.71°$ [1]

Framework density: 20.6 T/1000Å^3

Channels: [001] **8** 3.6 x 3.9*

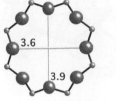 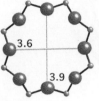

8-ring viewed along [001]

References:
(1) Barrett, P.A., Diaz-Cabanas, M.-J. and Camblor, M.A. *Chem. Mater.*, **11**, 2919–2927 (1999)
(2) Plévert, J., Yamamoto, K. Chiari, G. and Tatsumi, T. *J. Phys. Chem. B*, **103**, 8647–8649 (1999)

MTN

Framework Type Data

$Fd\overline{3}m$

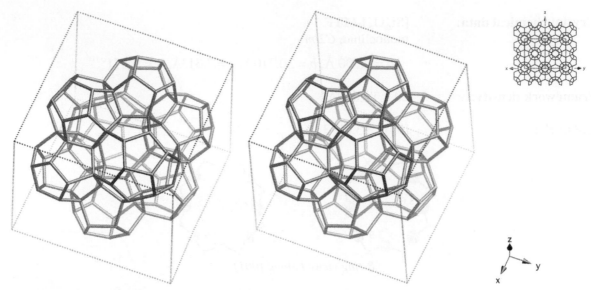

framework viewed along [111] (upper right: projection down [110])

Idealized cell data: cubic, $Fd\overline{3}m$ (origin choice 2), $a = 19.9$Å

Coordination sequences and vertex symbols:

T_1 (96,m)	4	12	25	43	68	95	133	177	223	274	5·5·5·5·5·6
T_2 (32,3m)	4	12	24	39	66	103	130	168	216	274	5·5·5·5·5·5
T_3 (8, $\overline{4}3m$)	4	12	24	36	64	112	132	156	222	264	5·5·5·5·5·5

Secondary building units: see *Compendium*

Composite building units:

mtn

Materials with this framework type:
*ZSM-39[1]
CF-4[2]
Dodecasil-3C[3]
Holdstite[4]

Crystal chemical data: $|(C_8H_{20}N)_q \,(OH)_q|\,[Si_{136}O_{272}]$-**MTN**
$C_8H_{20}N$ = tetraethylammonium
cubic, $Fd\overline{3}m$, $a = 19.36$Å $^{(1)}$

Framework density: 18.7 T/1000Å^3

Channels: apertures formed by 6-rings only

References:
(1) Schlenker, J.L., Dwyer, F.G., Jenkins, E.E., Rohrbaugh, W.J., Kokotailo G.T. and Meier, W.M. *Nature*, **294**, 340–342 (1981)
(2) Long, Y., He, H., Zheng, P., Guang, W. and Wang, B. *J. Incl. Phenom.*, **5**, 355–362 (1987)
(3) Gies, H. *Z. Kristallogr.*, **167**, 73–82 (1984)
(4) Smith, J.V. and Blackwell, C.S. *Nature*, **303**, 223–225 (1983)

MTT

Framework Type Data

Pmmn

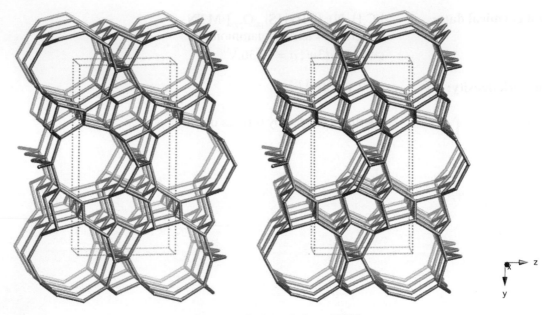

framework viewed along [100]

Idealized cell data: orthorhombic, *Pmmn* (origin choice 1), $a = 5.3$Å, $b = 22.0$Å, $c = 11.4$Å

Coordination sequences and vertex symbols:

T_1 (4,*m*)	4	12	23	43	66	92	130	170	213	261	$5 \cdot 5 \cdot 5 \cdot 5 \cdot 6 \cdot 10_2$
T_2 (4,*m*)	4	12	24	40	64	96	136	167	207	258	$5_2 \cdot 6_2 \cdot 6 \cdot 6_2 \cdot 6 \cdot 6_2$
T_3 (4,*m*)	4	12	22	40	67	97	124	165	219	265	$5 \cdot 5 \cdot 5 \cdot 5 \cdot 6_2 \cdot 10_2$
T_4 (4,*m*)	4	12	22	40	65	98	132	159	206	278	$5 \cdot 5 \cdot 5 \cdot 5 \cdot 6_2 \cdot *$
T_5 (4,*m*)	4	12	24	41	62	97	129	170	212	262	$5_2 \cdot 6_2 \cdot 6 \cdot 6_2 \cdot 6 \cdot 6_2$
T_6 (2,*mm2*)	4	12	22	42	66	94	126	164	220	270	$5 \cdot 5 \cdot 5 \cdot 5 \cdot 6_2 \cdot 10_2$
T_7 (2,*mm2*)	4	12	24	44	66	88	132	174	214	258	$5 \cdot 5 \cdot 5 \cdot 5 \cdot 10_2 \cdot *$

Secondary building units: 5-1

Composite building units:

jbw	mtt	bik	ton

Materials with this framework type:

*ZSM-23[1-4]   ISI-4[6]

EU-13[5]   KZ-1[7]

ZSM-23   **Type Material Data**   **MTT**

Crystal chemical data: $|Na_n (H_2O)_4| [Al_n Si_{24-n} O_{48}]$-**MTT**, n < 2
orthorhombic, $Pmn2_1$, $a = 21.5$Å, $b = 11.1$Å, $c = 5.0$ Å [(1)]

Framework density: 20.1 T/1000Å^3

Channels: [001] **10** 4.5 x 5.2*

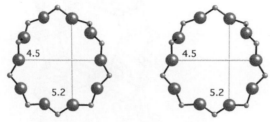

10-ring viewed along [001]

References:
(1) Schlenker, J.L., Higgins, J.B. and Cox, D.E. *private communication*
(2) Rohrman Jr., A.C., LaPierre, R.B., Schlenker, J.L., Wood, J.D., Valyocsik, E.W., Rubin, M.K., Higgins, J.B. and Rohrbaugh, W.J. *Zeolites*, **5**, 352–354 (1985)
(3) Marler, B., Deroche, C., Gies, H., Fyfe, C.A., Grondey, H., Kokotailo, G.T., Feng, Y., Ernst, S., Weitkamp, J. and Cox, D.E. *J. Appl. Crystallogr.*, **26**, 636–644 (1993)
(4) Zones, S.I., Darton, R.J., Morris, R. and Hwang, S.-J. *J. Phys. Chem. B*, **109**, 652–661 (2005)
(5) Araya, A. and Lowe, B.M. *U.S. Patent 4,581,211* (1986)
(6) Kakatsu, K. and Kawata, N. *Eur. Pat. Appl. EPA 102,497* (1984)
(7) Parker, L.M. and Bibby, D.M. *Zeolites*, **3**, 8–11 (1983)

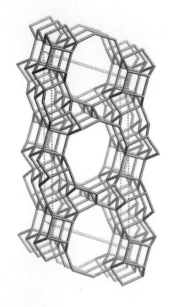
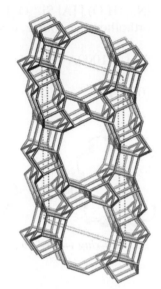

framework viewed along [010]

Idealized cell data: monoclinic, $C2/m$, $a = 25.6$Å, $b = 5.3$Å, $c = 12.1$Å, $\beta = 109.3°$

Coordination sequences and vertex symbols:

T_1 (4,*m*)	4	11	22	38	60	88	113	147	190	243	$4 \cdot 6_2 \cdot 5 \cdot 6 \cdot 5 \cdot 6$
T_2 (4,*m*)	4	11	22	38	60	86	115	147	191	238	$4 \cdot 6_2 \cdot 5 \cdot 6 \cdot 5 \cdot 6$
T_3 (4,*m*)	4	12	21	37	62	84	119	147	188	239	$5 \cdot 6 \cdot 5 \cdot 6 \cdot 5_2 \cdot 6$
T_4 (4,*m*)	4	12	23	37	59	85	120	154	184	231	$5 \cdot 6 \cdot 5 \cdot 6 \cdot 6 \cdot 6_2$
T_5 (4,*m*)	4	12	21	37	58	87	119	154	182	227	$5 \cdot 6_2 \cdot 5 \cdot 6_2 \cdot 5_2 \cdot 6$
T_6 (4,*m*)	4	12	24	39	55	85	122	156	188	225	$5 \cdot 6_2 \cdot 5 \cdot 6_2 \cdot 6_2 \cdot 12_6$
T_7 (4,*m*)	4	12	23	38	59	83	115	155	192	233	$5 \cdot 5 \cdot 5 \cdot 5 \cdot 6 \cdot 12_6$

Secondary building units: 5-[1,1]

Composite building units:

jbw	*cas*	*bik*	*mtw*

Materials with this framework type:

*ZSM-12[1,2]   CZH-5[5]   Theta-3[8]
[B-Si-O]-**MTW**[3]   NU-13[6]   VS-12[9]
[Ga-Si-O]-**MTW**[4]   TPZ-12[7]

234

Crystal chemical data: $|Na_n (H_2O)_8| [Al_n Si_{56-n} O_{112}]$-**MTW**, n < 5
monoclinic, $C2/c$

$a = 24.863Å, b = 5.012Å, c = 24.328Å, \beta = 107.72°^{(2)}$

Framework density: $19.4\ T/1000Å^3$

Channels: [010] **12** 5.6 x 6.0*

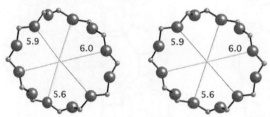

12-ring viewed along [010]

References:
(1) LaPierre, R.B., Rohrman Jr., A.C., Schlenker, J.L., Wood, J.D., Rubin, M.K. and Rohrbaugh, W.J. *Zeolites*, **5**, 346–348 (1985)
(2) Fyfe, C.A., Gies, H., Kokotailo, G.T., Marler, B. and Cox, D.E. *J. Phys. Chem.*, **94**, 3718–3721 (1990)
(3) Bandyopadhyay, R., Kubota, Y., Sugimoto, N., Fukushima, Y. and Sugi, Y. *Microporous Mesoporous Mat.*, **32**, 81–91 (1999)
(4) Zhi, Y.X., Tuel, A., Ben Taarit, Y. and Naccache, C. *Zeolites*, **12**, 138–141 (1992)
(5) Hickson, D.A. *UK Pat. Appl. GB 2079735A* (1981)
(6) Whittam, T.V. *Eur. Pat. Appl. EPA 0059059* (1982)
(7) Sumitani, K., Sakai, T., Yamasaki, Y. and Onodera, T. *U.S. Patent 4,557,919* (1985)
(8) Barlow, T.M. *E. Patent A-162,719* (1985)
(9) Reddy, K.M., Moudrakovski, I. and Sayari, A. *Chem. Commun.*, 1491–1492 (1994)

MWW

Framework Type Data

P6/mmm

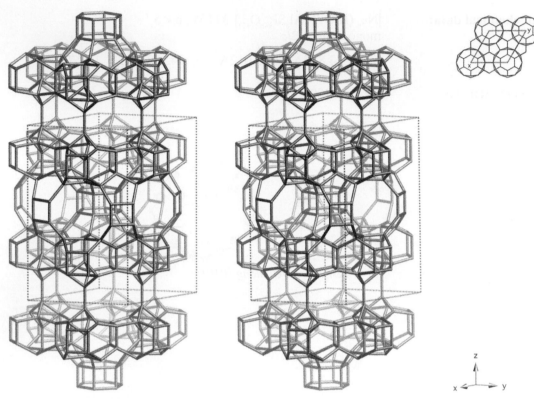

framework viewed normal to [001] (upper right: projection down [001])

Idealized cell data: hexagonal, *P6/mmm*, $a = 14.4$Å, $c = 25.2$Å

Coordination sequences and vertex symbols:
   see Appendix A for a list of the coordination sequences and vertex symbols for the 8 T-atoms

Secondary building units: see *Compendium*

Composite building units:
   d6r   *mel*

Materials with this framework type:
   *MCM-22[1]   [Ti-Si-O]-MWW[3]   ITQ-1[5,6]   SSZ-25[8,9]
   [Ga-Si-O]-MWW[2]   ERB-1[4]   PSH-3[7]

Crystal chemical data: $|H_{2.4}Na_{3.1}|\ [Al_{0.4}B_{5.1}Si_{66.5}O_{144}]$-**MWW**
hexagonal, $P6/mmm$, $a = 14.208$Å, $c = 24.945$Å [6]

Framework density: $16.5\ T/1000$Å^3

Channels: $\perp$ [001] **10** 4.0 x 5.5** | $\perp$ [001] **10** 4.1 x 5.1**

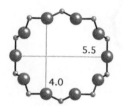
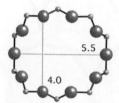
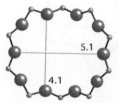
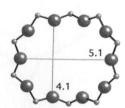

10-ring viewed normal to [001]   *10-ring viewed normal to [001]*
between 'layers'   *within 'layers'*

References:
(1) Leonowicz, M.E., Lawton, J.A., Lawton, S.L. and Rubin, M.K. *Science*, **264**, 1910–1913 (1994)
(2) Kim, S.J., Jung, K.D. and Joo, O.S. *J. Porous Mater.*, **11**, 211–218 (2004)
(3) Peng, W., Liu, Y.M., He, M.Y. and Tatsumi, T. *J. Catal.*, **228**, 183–191 (2004)
(4) Belussi, G., Perego, G., Clerici, M.G. and Giusti, A. *Eur. Pat. Appl. EPA 293032* (1988)
(5) Camblor, M.A., Corell, C., Corma, A., Diaz-Cabanas, M.-J., Nicolopoulos, S., Gonzalez-Calbet, J.M. and Vallet-Regi, M. *Chem. Mater.*, **8**, 2415–2417 (1996)
(6) Camblor, M.A., Corma, A., Diaz-Cabanas, M.-J. and Baerlocher, Ch. *J. Phys. Chem. B*, **102**, 44–51 (1998)
(7) Puppe, L. and Weisser, J. *U.S. Patent 4,439,409* (1984)
(8) Zones, S.I. *E. Patent 231,860* (1987)
(9) Zones, S.I., Hwang, S.J. and Davis, M.E. *Chemistry - A European Journal*, **7**, 1990–2001 (2001)

NAB

Framework Type Data

$I\bar{4}m2$

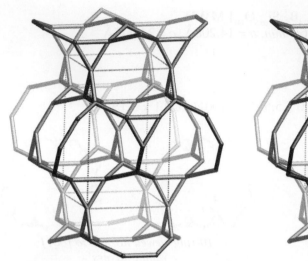 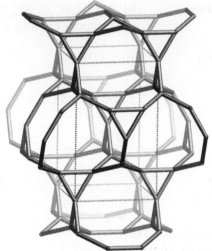

framework viewed along [100]

Idealized cell data: tetragonal, $I\bar{4}m2$, $a = 7.2$Å, $c = 12.0$Å

Coordination sequences and vertex symbols:

T_1 (8,.m.)	4	9	19	40	55	80	115	138	183	229	260	328	$3 \cdot 4 \cdot 8 \cdot 9_4 \cdot 8 \cdot 9_4$
T_2 (2,$\bar{4}m2$)	4	8	20	40	52	82	116	132	184	236	252	322	$3 \cdot 3 \cdot 9_4 \cdot 9_4 \cdot 9_4 \cdot 9_4$

Secondary building units: 4-1

Composite building units:

    lov   *vsv*

Materials with this framework type:

    *Nabesite[1]

Nabesite   **Type Material Data**   **NAB**

Crystal chemical data: $|Na_8 (H_2O)_{16}| [Be_4Si_{16}O_{40}]$-**NAB**
orthorhombic, $P2_12_12_1$, $a = 9.748$Å, $b = 10.133$Å, $c = 11.954$Å [1]
(Relationship to unit cell of Framework Type:

$a' = a\sqrt{2}$, $b' = b\sqrt{2}$, $c' = c$
or, as vectors, $\mathbf{a}' = \mathbf{a} + \mathbf{b}$, $\mathbf{b}' = \mathbf{b} - \mathbf{a}$, $\mathbf{c}' = \mathbf{c}$)

Stability: reversable dehydration [1]

Framework density: 16.9 T/1000Å^3

Channels: [110] **9** 2.7 x 4.1* $\leftrightarrow$ [$\overline{1}\overline{1}$0] **9** 3.0 x 4.6*

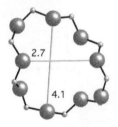
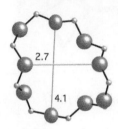

9-ring viewed along [110]

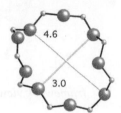
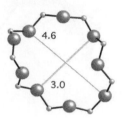

9-ring viewed along [$\overline{1}\overline{1}$0]

References:
(1)   Petersen, O.V., Giester, G., Brandstätter, F. and Niedermayr, G. *Can. Mineral.*, **40**, 173–181 (2002)

NAT

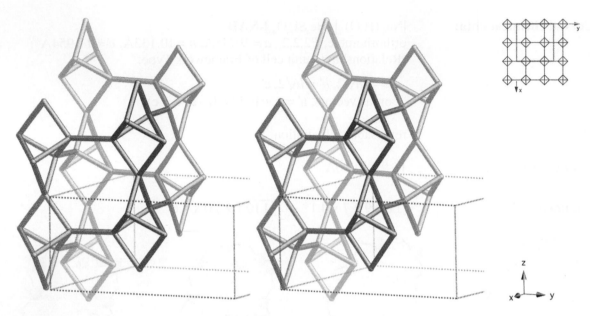

framework viewed normal to [001]

Idealized cell data: tetragonal, $I4_1/amd$ (origin choice 2), $a = 13.9$, $c = 6.4$

Coordination sequences and vertex symbols:

T_1 (16,*m*)	4	9	19	35	52	78	106	139	179	213
T_2 (4,$\overline{4}m2$)	4	8	18	36	56	66	116	140	154	232

$4 \cdot 8_2 \cdot 4 \cdot 8_2 \cdot 4_2 \cdot 8_4$

$4_2 \cdot 4_2 \cdot 8_4 \cdot 8_4 \cdot 8_4 \cdot 8_4$

Secondary building units: 4=1

Composite building units:

    nat

Materials with this framework type:

*Natrolite[1,2]
[Al-Ge-O]-**NAT**[3]
[Ga-Si-O]-**NAT**[4]
|Rb-|[Ga-Ge-O]-**NAT**[5]
Gonnardite[6]

High natrolite[7]
Mesolite[8,9]
Metanatrolite[10]
Paranatrolite[11]
Scolecite[8,12-14]

Synthetic gonnardite[15]
Synthetic mesolite[16]
Synthetic natrolite[16]
Synthetic scolecite[16]
Tetranatrolite[17]

Natrolite

Type Material Data

NAT

Crystal chemical data: $|Na_{16}(H_2O)_{16}|$ $[Al_{16}Si_{24}O_{80}]$-**NAT**
orthorhombic, $Fdd2$, $a = 18.30$Å, $b = 18.63$Å, $c = 6.60$Å [2]
(Relationship to unit cell of Framework Type:

$$a' = b' = a\sqrt{2}, c' = c$$
or, as vectors, $\mathbf{a'} = \mathbf{a} + \mathbf{b}, \mathbf{b'} = \mathbf{b} - \mathbf{a}, \mathbf{c'} = \mathbf{c})$

Framework density: 17.8 T/1000Å^3

Channels: <100> **8** 2.6 x 3.9** $\leftrightarrow$ [001] **9** 2.5 x 4.1*
(variable due to considerable flexibility of framework)

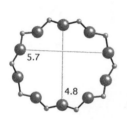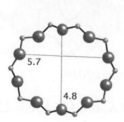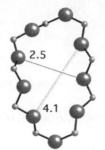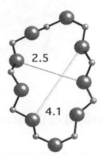

8-ring viewed along <100>   *9-ring viewed along [001]*

References:
(1) Pauling, L. *Proc. Natl. Acad. Sci.*, **16**, 453–459 (1930)
(2) Meier, W.M. *Z. Kristallogr.*, **113**, 430–444 (1960)
(3) Tripathi, A., Johnson, G.M., Kim, S.J. and Parise, J.B. *J. Mater. Chem.*, **10**, 451–455 (2000)
(4) Xie, D., Newsam, J.M., Yang, J. and Yelong, W.B. *MRS Sym. Proc.*, **111**, pp. 147–154 (1988)
(5) Klaska, K.H. and Jarchow, O. *Z. Kristallogr.*, **172**, 167–174 (1985)
(6) Mazzi, F., Larsen, A.O., Gottardi, G. and Galli, E. *N. Jb. Miner. Mh.*, 219–228 (1986)
(7) Baur, W.H. and Joswig, W. *N. Jb. Miner. Mh.*, 171–187 (1996)
(8) Taylor, W.H., Meek, C.A. and Jackson, W.W. *Z. Kristallogr.*, **84**, 373–398 (1933)
(9) Artioli, G., Smith, J.V. and Pluth, J.J. *Acta Crystallogr.*, **C42**, 937–942 (1986)
(10) Joswig, W. and Baur, W.H. *N. Jb. Miner. Mh.*, 26–38 (1995)
(11) Seryotkin, Y.V., Bakakin, V.V. and Belitsky, I.A. *Eur. J. Mineral.*, **16**, 545–550 (2004)
(12) Fälth, L. and Hansen, S. *Acta Crystallogr.*, **B35**, 1877–1880 (1979)
(13) Smith, J.V., Pluth, J.J., Artioli, G. and Ross, F.K. *Proc. 6th Int. Zeolite Conf.*, pp. 842–850 (1984)
(14) Stuckenschmidt, E., Joswig, W., Baur, W.H. and Hofmeister, W. *Phys. Chem. Mineral.*, **24**, 403–410 (1997)
(15) Ghobarkar, H. and Schaef, O. *Zeolites*, **19**, 259–261 (1997)
(16) Ghobarkar, H. and Schaef, O. *Cryst. Res. Technol.*, **31**, K67–69 (1996)
(17) Evans, H.T., Konnert, J.A., Ross, M. *Am. Mineral.*, **85**, 1808–1815 (2000)

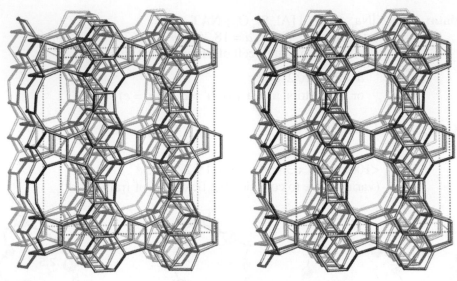

framework viewed along [010]

Idealized cell data: orthorhombic, *Fmmm*, $a = 26.1\text{Å}$, $b = 13.9\text{Å}$, $c = 22.9\text{Å}$

Coordination sequences and vertex symbols:

T_1 (32,1)	4	11	21	36	58	89	123	157	187	237	$4 \cdot 6 \cdot 5 \cdot 5_2 \cdot 10$
T_2 (32,1)	4	12	22	38	57	86	118	152	196	245	$5 \cdot 5 \cdot 5 \cdot 6_2 \cdot 5 \cdot 10$
T_3 (16,*m*)	4	12	20	34	57	88	125	158	192	224	$5 \cdot 5_2 \cdot 5 \cdot 5_2 \cdot 12_2 \cdot *$
T_4 (16,*m*)	4	12	20	31	57	84	118	150	187	237	$5 \cdot 5 \cdot 5 \cdot 5 \cdot 5 \cdot 6_2$
T_5 (16,*m*)	4	11	24	40	63	86	114	158	208	255	$4 \cdot 10 \cdot 5 \cdot 5 \cdot 5 \cdot 5$
T_6 (16,*m*)	4	12	22	35	55	83	119	151	184	237	$5 \cdot 5_2 \cdot 5 \cdot 6 \cdot 5 \cdot 6$
T_7 (8,*mm*2)	4	12	24	32	50	88	120	152	180	226	$5 \cdot 5 \cdot 5 \cdot 5 \cdot 12_6 \cdot *$

Secondary building units: see *Compendium*

Composite building units:

    cas   *non*   *ton*

Materials with this framework type:
  *NU-87[1]
  Gottardiite[2]

Type Material: NU-87   Type Material Data

Crystal chemical data: |H$_4$ (H$_2$O)$_n$| [Al$_4$Si$_{64}$O$_{136}$]-**NES**
monoclinic, $P2_1/c$

a = 14.324Å, b = 22.376Å, c = 25.092Å, β = 151.51° [1]
(Relationship to unit cell of Framework Type:

a' = $b/(2\sin\beta')$, b' = c, c' = a
or, as vectors, **a'** = (**b** - **a**)/2, **b'** = **c**, **c'** = **a**)

Framework density: 17.7 T/1000Å^3

Channels: [100] **10** 4.8 x 5.7**

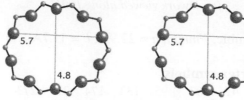

10-ring along [100]

References:
(1) Shannon, M.D., Casci, J.L., Cox, P.A. and Andrews, S.J. *Nature*, **353**, 417–420 (1991)
(2) Alberti, A., Vezzalini, G., Galli, E. and Quartieri, S. *Eur. J. Mineral.*, **8**, 69–75 (1996)

NON

Fmmm

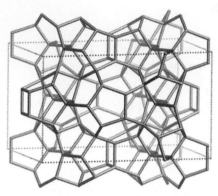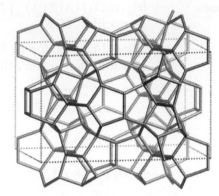

framework viewed along [001]

Idealized cell data: orthorhombic, *Fmmm*, $a = 22.9$Å, $b = 15.7$Å, $c = 13.9$Å

Coordination sequences and vertex symbols:

T_1 (32,1)	4	12	25	42	67	95	133	174	219	273	$5 \cdot 6 \cdot 5 \cdot 6 \cdot 5 \cdot 6$
T_2 (16,*m*)	4	12	24	39	64	99	130	174	217	262	$5 \cdot 6 \cdot 5 \cdot 6 \cdot 5 \cdot 6_2$
T_3 (16,*m*)	4	11	23	44	72	95	124	170	229	279	$4 \cdot 6 \cdot 5 \cdot 5 \cdot 5 \cdot 5$
T_4 (16,*m*)	4	12	24	41	65	97	133	173	212	267	$5 \cdot 5_2 \cdot 5 \cdot 6 \cdot 5 \cdot 6$
T_5 (8,2*mm*)	4	12	24	40	62	92	142	166	214	262	$5 \cdot 5 \cdot 5 \cdot 5 \cdot 12_2 \cdot *$

Secondary building units: see *Compendium*

Composite building units:

non

Materials with this framework type:
*Nonasil[1]
[B-Si-O]-**NON**[2]
|(Co(C_5H_5)_2)_4 F_4|[Si_88O_176]-**NON**[3]
CF-3[4]
ZSM-51[5]

Nonasil   **Type Material Data**   **NON**

Crystal chemical data: | $(C_5H_{13}N)_4$ | $[Si_{88}O_{176}]$-**NON**
                              $C_5H_{13}N$ = 2-aminopentane
                              orthorhombic, *Fmmm*, $a = 22.232$Å, $b = 15.058$Å, $c = 13.627$Å [1]

Framework density: 19.3 T/1000Å^3

Channels: apertures formed by 6-rings only

References:
(1) Marler, B., Dehnbostel, N., Eulert, H.-H., Gies, H. and Liebau, F. *J. Incl. Phenom.*, **4**, 339–349 (1986)
(2) Marler, B. and Gies, H. *Zeolites*, **15**, 517–525 (1995)
(3) Vandegoor, G., Freyhardt, C.C. and Behrens, P. *Z. anorg. allg. Chemie*, **621**, 311–322 (1999)
(4) Long, Y.-C., Zhong, W. and Shen, X. *J. Incl. Phenom.*, **4**, 121–127 (1986)
(5) Rohrbaugh, W.J. *private communication*

NPO

Framework Type Data

P6₃/mmc

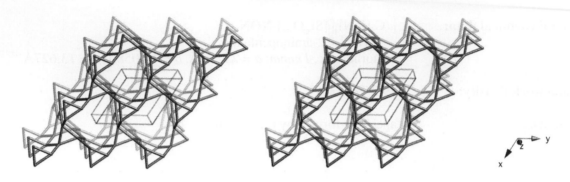

framework viewed along [001]

Idealized cell data: hexagonal, P6₃/mmc, $a = 9.1$Å, $c = 5.3$Å

Coordination sequences and vertex symbols:

T₁ (6,mm2)	4	10	20	34	58	82	108	144	186	222	268	330		3·6₂·6·6·6·6

Secondary building units: 3

Materials with this framework type:
*Nitridophosphate-1[1]

Nitridophosphate-1   **Type Material Data**   **NPO**

Crystal chemical data: $|Li_xH_{12-x-y+z}Cl_z|$ $[P_{12}O_yN_{24-y}]$-**NPO**
     with $6 < x < 9$, $2 < y < 4$ and $2 < z < 3$
     orthorhombic, $Pna2_1$, $a = 4.753$Å, $b = 14.208$Å, $c = 8.203$Å [1]
     (Relationship to unit cell of Framework Type:

     $a' = c, b' = a\sqrt{3}$, $c' = b$
     or, as vectors, $\mathbf{a'} = \mathbf{c}, \mathbf{b'} = 2\mathbf{a} + \mathbf{b}, \mathbf{c'} = \mathbf{b}$)

Framework density: 21.7 T/1000Å³

Channels: [100] **12** 3.3 x 4.4*

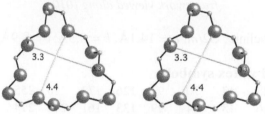

12-ring viewed along [100]

References:
(1) Correll, S., Oeckler, O., Stock, N. and Schnick, W. *Angew. Chem., Int. Ed.*, **42**, 3549–3552 (2003)

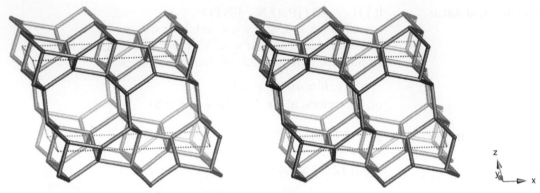

framework viewed along [010]

Idealized cell data: monoclinic, *C2/m*, *a* = 14.1Å, *b* = 5.3Å, *c* = 8.9Å, β = 105.4°

Coordination sequences and vertex symbols:

T_1 (4,*m*)	4	12	23	42	70	93	126	172	216	255	314	385	$5 \cdot 6 \cdot 5 \cdot 6 \cdot 5_2 \cdot 6$
T_2 (4,*m*)	4	12	26	43	64	99	133	161	210	274	318	364	$5 \cdot 6 \cdot 5 \cdot 6 \cdot 6_2 \cdot 8_2$
T_3 (4,*m*)	4	12	23	40	68	95	123	169	217	256	310	383	$5 \cdot 5 \cdot 5 \cdot 5 \cdot 6 \cdot 8_2$

Secondary building units: 5-1

Composite building units:

      cas   *bik*

Materials with this framework type:
*Nu-6(2)[1]
EU-20b (**CAS-NSI** structural intermediate)[2]

Nu-6(2)   **Type Material Data**   **NSI**

Crystal chemical data: $[Si_{24}O_{48}]$-**NSI**
monoclinic, $P2_1/a$

$a = 17.257$Å, $b = 4.988$Å, $c = 13.848$Å, $\beta = 106.1°$ [1]
(Relationship to unit cell of Framework Type: $a' = 2c$, $b' = b$, $c' = a$)

Framework density: 21 T/1000Å³

Channels: [010] **8** 2.6 x 4.5* | [010] **8** 2.4 x 4.8*

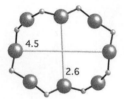 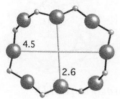 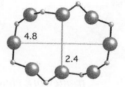 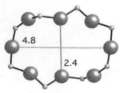

8-ring viewed along [010]   *2nd 8-ring viewed along [010]*

References:
(1) Zanardi, S., Alberti, A., Cruciani, G., Corma, A., Fornés, V. and Brunelli, M. *Angew. Chem., Int. Ed.*, **43**, 4933–4937 (2004)
(2) Marler, B., Camblor, M.A. and Gies, H. *Microporous Mesoporous Mat.*, **90**, 87–101 (2006)

OBW

Framework Type Data

I4/mmm

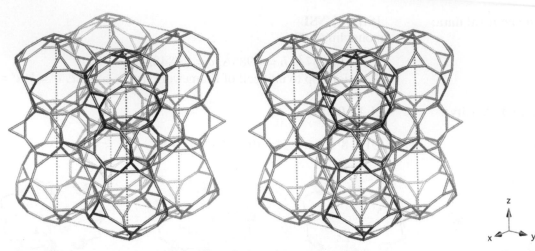

framework viewed normal to [001]

Idealized cell data: tetragonal, *I4/mmm*, a = 13.9Å, c = 30.8Å

Coordination sequences and vertex symbols:

T₁ (32,1)	4	8	16	30	45	69	95	126	162	194	242	295	3·3·8·8·8·10
T₂ (16,.m.)	4	8	16	30	50	67	93	122	154	204	256	292	3·3·8·10·8·10
T₃ (16,..m)	4	8	17	30	43	69	97	124	174	195	229	305	3·3·8·8·8·8
T₄ (8,m.2m)	4	9	18	32	46	64	93	130	170	214	247	260	3·4·8·8·8·8
T₅ (4,4̄m2)	4	8	16	32	52	72	90	120	160	208	264	320	3·3·10·10·10·10

Secondary building units: see *Compendium*

Composite building units:

lov   *vsv*

Materials with this framework type:
   *OSB-2[1,2]

OSB-2 **Type Material Data** **OBW**

Crystal chemical data: $|K_{44} (H_2O)_{96}| [Be_{22} Si_{54}O_{150}]$-**OBW**
tetragonal, $I4/mmm$, $a = 13.7452$Å, $c = 30.654$Å [2]

Stability: NH_4-exchanged form not stable to heating [2]

Framework density: 13.1 T/1000Å³

Channels: {<110> **10** 5.0 x 5.0** ↔ ([001] **8** 3.4 x 3.4* + <101> **8** 2.8 x 4.0**)

↔ <100> **8** 3.3 x 3.4**}***

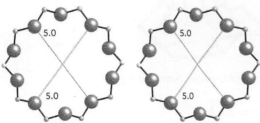
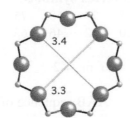

10-ring viewed along [110]   *8-ring viewed along [001]*

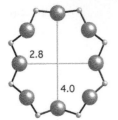
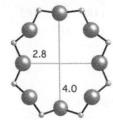

8-ring viewed along [101]   *8-ring viewed along [100]*

References:
(1) Cheetham, A.K., Fjellvåg, H., Gier, T.E., Kongshaug, K.O., Lillerud, K.P. and Stucky, G.D. *Stud. Surf. Sci. Catal.*, **135**, 158 (2001)
(2) Lillerud, K.P. *private communication*

OFF

Framework Type Data

$P\bar{6}m2$

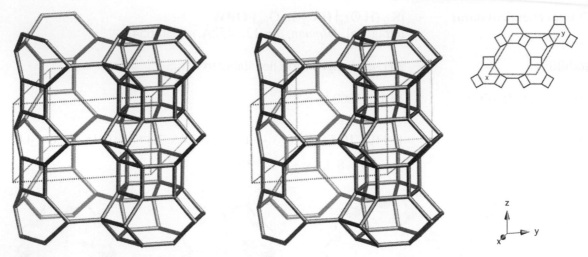

framework viewed normal to [001] (upper right: projection down [001])

Idealized cell data: hexagonal, $P\bar{6}m2$, $a = 13.1$Å, $c = 7.6$Å

Coordination sequences and vertex symbols:

T_1 (12,1)	4	9	17	30	50	75	98	118	144	185	4·4·4·6·6·8
T_2 (6,m)	4	10	20	32	46	66	94	128	162	192	4·8·4·8·6·6

Secondary building units: 6 or 4-2

Framework description: AAB sequence of 6-rings

Composite building units:

dsc	*d6r*	*can*	*gme*
double sawtooth chain			

Materials with this framework type:

*Offretite[1-4]

LZ-217[5]

Linde T (**ERI-OFF** structural intermediate)[6]

Synthetic offretite[7]

TMA-O[8]

Offretite

Type Material Data

OFF

Crystal chemical data: $|(Ca,Mg)_{1.5}K(H_2O)_{14}|\ [Al_4Si_{14}O_{36}]$-**OFF**
hexagonal, $P\bar{6}m2$, $a = 13.291$Å, $c = 7.582$Å $^{(2)}$

Framework density: 15.5 T/1000Å^3

Channels: [001] **12** 6.7 x 6.8* ↔ ⊥ [001] **8** 3.6 x 4.9**

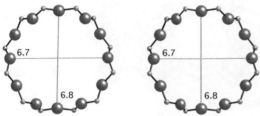
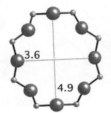
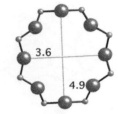

12-ring viewed along [001]   *8-ring viewed normal to [001]*

References:
(1) Bennett, J.M. and Gard, J.A. *Nature*, **214**, 1005–1006 (1967)
(2) Gard, J.A. and Tait, J.M. *Acta Crystallogr.*, **B28**, 825–834 (1972)
(3) Mortier, W.J., Pluth, J.J. and Smith, J.V. *Z. Kristallogr.* **143**, 319–332 (1976)
(4) Alberti, A., Cruciani, G., Galli, E. and Vezzalini, G. *Zeolites*, **17**, 457–461 (1996)
(5) Breck, D.W. and Skeels, G.W. *U.S. Patent 4,503,023* (1985)
(6) Breck, D.W. *Zeolite Molecular Sieves*, p. 173 (1974)
(7) Ghobarkar, H. and Schaef, O. *Cryst. Res. Technol.*, **31**, K29-31 (1996)
(8) Aiello, R., Barrer, R.M., Davies, J.A. and Kerr, I.S. *Trans. Faraday Soc.*, **66**, 1610–1617 (1970)

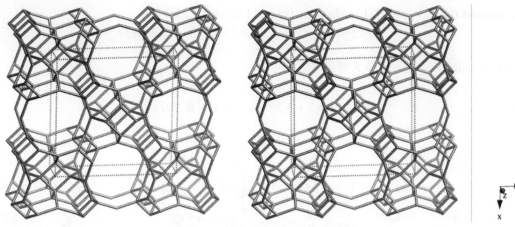

framework viewed along [001]

Idealized cell data: tetragonal, *I4/mmm*, $a = 18.5$Å, $c = 5.3$Å

Coordination sequences and vertex symbols:

T_1 (16,*m*)	4	10	21	37	57	82	111	145	189	236	$4 \cdot 6 \cdot 4 \cdot 6 \cdot 6 \cdot 6_2$
T_2 (8,*m2m*)	4	11	22	34	52	84	120	149	180	220	$4 \cdot 6_2 \cdot 6_2 \cdot 6_2 \cdot 6_2 \cdot 6_2$
T_3 (8,*m2m*)	4	12	22	37	60	81	112	154	192	230	$6_2 \cdot 6_2 \cdot 6_2 \cdot 6_2 \cdot 6_2 \cdot 12_6$

Secondary building units: 6-2

Composite building units:

dzc	*ats*	*lau*
double zigzag chain		

Materials with this framework type:
  *UiO-6[1]

Crystal chemical data: $[Al_{16}P_{16}O_{64}]$-**OSI**
orthorhombic, *Imm*2, $a = 18.355Å$, $b = 18.321Å$, $c = 5.053Å$ [1]

Framework density: $18.8 \text{ T}/1000Å^3$

Channels: [001] **12** 5.2 x 6.0*

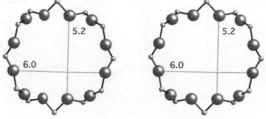

12-ring viewed along [001]

References:
(1) Akporiaye, D.E., Fjellvag, H., Halvorsen, E.N., Haug, T., Karlsson, A. and Lillerud, K.P. *Chem. Commun.*, 1553–1554 (1996)

OSO

$P6_222$

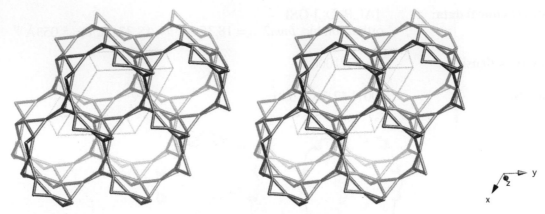

framework viewed along [001]

Idealized cell data: hexagonal, $P6_222$, $a = 10.1$Å, $c = 7.6$Å

Coordination sequences and vertex symbols:

T_1 (6,2)	4	8	16	29	46	70	101	118	162	190	$3·3·8·8·8·14_{10}$
T_2 (3,222)	4	8	16	30	44	76	92	130	148	202	$3·3·8·8·14_7·14_7$

Secondary building units: see *Compendium*

Composite building units:
    lov

Materials with this framework type:
    *OSB-1[1,2]

OSB-1 **Type Material Data** **OSO**

Crystal chemical data: $|K_6 (H_2O)_9| [Be_3Si_6O_{18}]$-**OSO**
trigonal, $P3_2$, $a = 10.093Å$, $c = 7.626Å$ [(1)]

Framework density: $13.4 \ T/1000Å^3$

Channels: [001] **14** 5.4 x 7.3* $\leftrightarrow$ $\perp$ [001] **8** 2.8 x 3.3**

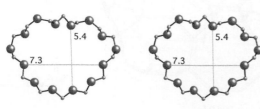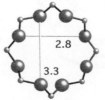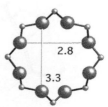

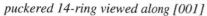

puckered 14-ring viewed along [001]   *8-ring viewed normal to [001]*

References:
(1) Kongshaug, K.O., Fjellvåg, H., Lillerud, K.P., Gier, T.E., Stucky, G.D. and Cheetham, A.K. *private communication*
(2) Cheetham, A.K., Fjellvåg, H., Gier, T.E., Kongshaug, K.O., Lillerud, K.P. and Stucky, G.D. *Stud. Surf. Sci. Catal.*, **135**, 158 (2001)

OWE

Framework Type Data

Pmma

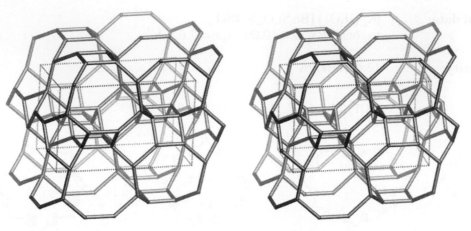

framework viewed along [010]

Idealized cell data: orthorhombic, *Pmma*, $a = 14.4$Å, $b = 7.2$Å, $c = 9.1$Å

Coordination sequences and vertex symbols:

T_1 (8,1)	4	9	18	35	57	76	96	128	172	216	252	290	$4 \cdot 4 \cdot 4 \cdot 8_3 \cdot 6_2 \cdot 8_2$
T_2 (4,.m.)	4	9	19	34	52	75	99	129	172	215	250	288	$4 \cdot 6_2 \cdot 4 \cdot 6_2 \cdot 4 \cdot 8_2$
T_3 (4,.m.)	4	11	22	32	49	77	109	139	163	197	250	310	$4 \cdot 8_2 \cdot 6_2 \cdot 8_2 \cdot 6_2 \cdot 8_2$

Secondary building units: 4 or 4-4-

Composite building units:

   dsc   *sti*

   double sawtooth
   chain

Materials with this framework type:
 *UiO-28[1]
 ACP-2[2]

UiO-28   **Type Material Data**   **OWE**

Crystal chemical data: $|(C_4N_3H_{14})_4 (H_2O)_4| [Mg_4Al_{12}P_{16}O_{64}]$-**OWE**
                $C_4N_3H_{13}$ = diethylenetriamine
         orthorhombic, *Pbcm*, a = 9.2769Å, b = 14.7984Å, c = 14.6106Å [1]
         (Relationship to unit cell of Framework Type: a' = c, b' = $2b$, c' = a)

Framework density: 16 T/1000Å^3

Channels: [010] **8** 3.5 x 4.0* $\leftrightarrow$ [001] **8** 3.2 x 4.8*

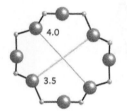 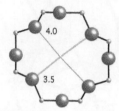 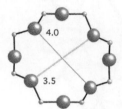 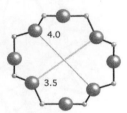

         8-ring viewed along [010]   *8-ring viewed along [001]*

References:
(1)   Kongshaug, K.O., Fjellvag, H. and Lillerud, K.P. *J. Mater. Chem.*, **11**, 1242–1247 (2001)
(2)   Feng, P., Bu, X. and Stucky, G.D. *Nature*, **388**, 735–741 (1997)

-PAR

Framework Type Data

C2/c

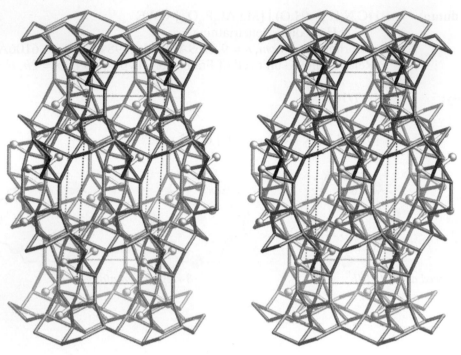

framework viewed along [001]

Idealized cell data: monoclinic, *C2/c*, $a = 20.9$Å, $b = 9.2$Å, $c = 8.6$Å, $\beta = 89.7°$

Coordination sequences and vertex symbols:

T_1 (8,1)	4	9	18	31	48	71	99	132	162	197		$4 \cdot 6 \cdot 6 \cdot 6 \cdot 6 \cdot 10_2$
T_2 (8,1)	4	9	19	33	51	77	96	126	162	203		$4 \cdot 6 \cdot 4 \cdot 8_2 \cdot 6 \cdot 6_2$
T_3 (8,1)	4	10	19	33	52	72	102	126	161	204		$4 \cdot 6_2 \cdot 4 \cdot 8_2 \cdot 8_2 \cdot 10$
T_4 (8,1)	3	8	16	29	49	68	94	123	162	203		$4 \cdot 6 \cdot 6_2$

Secondary building units: 4

Materials with this framework type:

*Partheite[1]

Partheite

Type Material Data

-PAR

Crystal chemical data: $|Ca_8 (H_2O)_{16}| [Al_{16}Si_{16}O_{60} (OH)_8]$-**PAR**
monoclinic, $C2/c$

$$a = 21.555\text{Å}, b = 8.761\text{Å}, c = 9.304\text{Å}, \beta = 91.55°^{(1)}$$

Stability: Stable at 150°C, transforms at 400°C[1]

Framework density: 18.2 T/1000Å^3

Channels: [001] **10** 3.5 x 6.9*

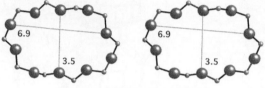

10-ring viewed along [001]

References:
(1)   Engel, N. and Yvon, K. Z. *Kristallogr.*, **169**, 165–175 (1984)

PAU

Framework Type Data

Im3̄m

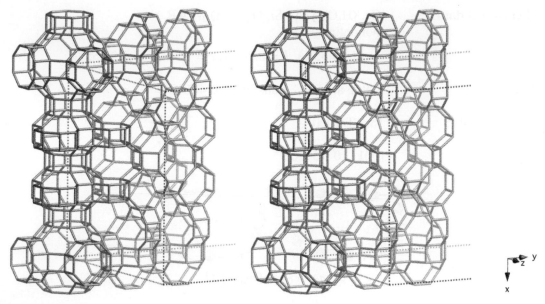

framework viewed along [001]

Idealized cell data: cubic, $Im\bar{3}m$, $a = 34.8$Å

Coordination sequences and vertex symbols:
see Appendix A for a list of the coordination sequences and vertex symbols for the 8 T-atoms

Secondary building units: 4 or 8

Composite building units:

d8r   pau   lta

Materials with this framework type:
*Paulingite[1,2]
[Ga-Si-O]-**PAU**[3]
ECR-18[4,5]
Paulingite, Vinaricka Hora[6]

Paulingite   **Type Material Data**   **PAU**

Crystal chemical data: | (Ca,K$_2$,Na$_2$)$_{76}$ (H$_2$O)$_{700}$| [Al$_{152}$Si$_{520}$O$_{1344}$]-**PAU**
cubic, $Im\overline{3}m$, $a = 35.093$Å $^{(1)}$

Framework density: 15.5 T/1000Å^3

Channels: <100> **8** 3.6x3.6*** | <100> **8** 3.6 x 3.6***

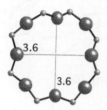 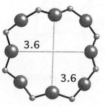

8-ring viewed along <100>

References:
(1) Gordon, E.K., Samson, S. and Kamb, W.K. *Science*, **154**, 1004–1007 (1966)
(2) Bieniok, A. *Natural Zeolites for the Third Millennium*, 53–60 (2000)
(3) Kim, D.J., Shin, C.H. and Hong, S.B. *Microporous Mesoporous Mat.*, **83**, 319–325 (2005)
(4) Vaughan, D.E.W. and Strohmaier, G. *U.S. Patent 4,661,332* (1987)
(5) Vaughan, D.E.W. and Strohmaier, K.G. *Microporous Mesoporous Mat.*, **28**, 233–239 (1999)
(6) Lengauer, C.L., Giester, G. and Tillmanns, E. *Mineral. Mag.*, **61**, 591–606 (1997)

PHI

Framework Type Data

Cmcm

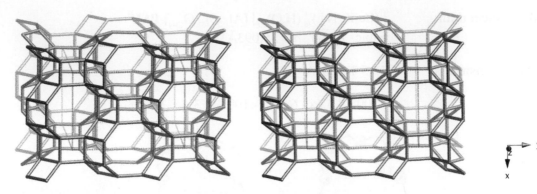

framework viewed along [001]

Idealized cell data: orthorhombic, *Cmcm*, $a = 9.9\text{Å}$, $b = 14.1\text{Å}$, $c = 14.0\text{Å}$

Coordination sequences and vertex symbols:

T_1 (16,1)	4	9	18	32	50	71	94	122	157	195	$4\cdot4\cdot4\cdot8_2\cdot8\cdot8$
T_2 (16,1)	4	9	18	32	48	68	96	126	155	191	$4\cdot4\cdot4\cdot8_2\cdot8\cdot8$

Secondary building units: 8 or 4

Composite building units:
> *dcc*
> *double*
> *crankshaft chain*

Materials with this framework type:

*Phillipsite[1,2]

[Al-Co-P-O]-**PHI**[3]

DAF-8[4]

Harmotome[2,5]

Wellsite[6]

ZK-19[7]

Phillipsite   **Type Material Data**   **PHI**

Crystal chemical data: $|K_2 (Ca,Na_2)_2 (H_2O)_{12}|$ $[Al_6Si_{10}O_{32}]$-**PHI**
monoclinic, $P2_1/m$

$a = 9.865$Å, $b = 14.300$Å, $c = 8.668$Å, $\beta = 124.20°$ [2]
(Relationship to unit cell of Framework Type:

$a' = a$, $b' = c$, $c' = b/(2\sin(\beta'))$
or, as vectors, **a' = a, b' = c, c' = (b - a)/2**)

Framework density: 15.8 T/1000Å^3

Channels: [100] **8** 3.8 x 3.8* $\leftrightarrow$ [010] **8** 3.0 x 4.3* $\leftrightarrow$ [001] **8** 3.2 x 3.3*

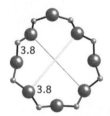 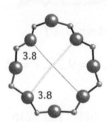 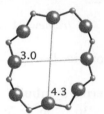 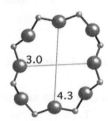

8-ring viewed along [100]   *8-ring viewed along [010]*

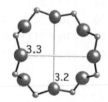 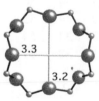

8-ring viewed along [001]

References:
(1) Steinfink, H. *Acta Crystallogr.*, **15**, 644–651 (1962)
(2) Rinaldi, R., Pluth, J.J. and Smith, J.V. *Acta Crystallogr.*, **B30**, 2426–2433 (1974)
(3) Feng, P., Bu, X. and Stucky, G.D. *Nature*, **388**, 735–741 (1997)
(4) Barrett, P.A., Sankar, G., Stephenson, R., Catlow, C.R.A., Thomas, J.M., Jones, R.H. and Teat, S.J. *Solid State Sci.*, **8**, 337–341 (2006)
(5) Sadanaga, R., Marumo, F. and Yakéuchi, Y. *Acta Crystallogr.*, **14**, 1153–1163 (1961)
(6) Cerny, P., Rinaldi, R. and Surdam, R.C. *N. Jb. Miner. Abh.*, **128**, 312–330 (1977)
(7) Kuehl, G.H. *Am. Mineral.*, **54**, 1607–1612 (1969)

PON

Framework Type Data

Pca2₁

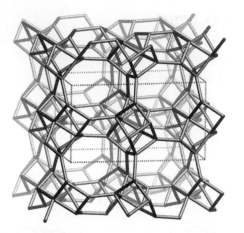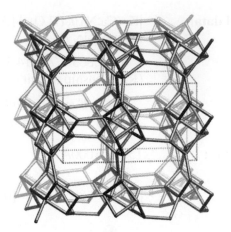

framework viewed along [100]

Idealized cell data: orthorhombic, $Pca2_1$, $a = 8.9$Å, $b = 9.2$Å, $c = 16.1$Å

Coordination sequences and vertex symbols:

T_1 (4,1)	4	10	20	36	56	76	105	144	180	215	260	323	$4·6·4·6·6·8$
T_2 (4,1)	4	10	20	34	54	82	108	134	175	224	264	304	$4·6·4·6·6_2·10_7$
T_3 (4,1)	4	10	19	34	56	78	106	144	179	211	262	321	$4·6·4·6_2·6_3·6_4$
T_4 (4,1)	4	11	20	34	55	83	110	136	176	227	264	304	$4·6·6·6_2·6_4·10_7$
T_5 (4,1)	4	9	18	34	54	75	104	142	176	211	260	319	$4·4·4·6·6·6_3$
T_6 (4,1)	4	10	21	38	57	76	105	146	182	212	262	325	$4·6·4·6·6_2·8$

Secondary building units: 4-2

Materials with this framework type:
    *IST-1[1]

IST-1   **Type Material Data**   **PON**

Crystal chemical data: |(CH$_3$NH$_3$)$_4$ (CH$_3$NH$_2$)$_4$ (OH)$_4$| [Al$_{12}$P$_{12}$O$_{48}$]-**PON**
orthorhombic, *Pca*2$_1$, *a* = 9.6152Å, *b* = 8.6702Å, *c* = 16.2196Å [1]

Framework density: 17.7 T/1000Å^3

Channels: [100] **10** 5.0 x 5.3*

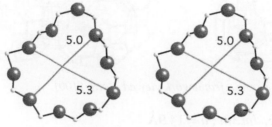

10-ring viewed along [100]

References:
(1) Jorda, J.L., McCusker, L.B., Baerlocher, Ch., Morais, C.M., Rocha, J., Fernandez, C., Borges, C., Lourenco, J.P., Ribeiro, M.F. and Gabelica, Z. *Microporous Mesoporous Mat.*, **65**, 43–57 (2003)

RHO

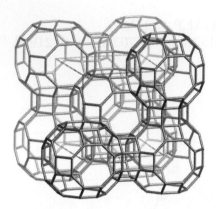 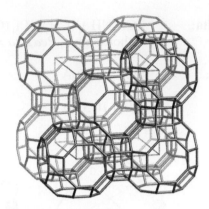

framework viewed along [100]

Idealized cell data: cubic, $Im\bar{3}m$, $a = 14.9$Å

Coordination sequences and vertex symbols:

| T_1 (48,2) | 4 | 9 | 17 | 28 | 42 | 60 | 81 | 105 | 132 | 162 | | 4·4·4·6·8·8 |

Secondary building units: 8-8 or 8 or 6 or 4

Composite building units:

d8r   *lta*

    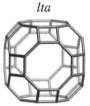

Materials with this framework type:

*Rho[1,2]
[Be-As-O]-**RHO**[3]
[Be-P-O]-**RHO**[4]
[Co-Al-P-O]-**RHO**[5]
[Mg-Al-P-O]-**RHO**[5]

[Mn-Al-P-O]-**RHO**[5]
|Na$_{16}$ Cs$_8$|[Al$_{24}$Ge$_{24}$O$_{96}$]-**RHO**[6]
Gallosilicate ECR-10[7]
LZ-214[8]
Pahasapaite[9,10]

Rho **Type Material Data** **RHO**

Crystal chemical data: $|(Na,Cs)_{12} (H_2O)_{44}| [Al_{12}Si_{36}O_{96}]$-**RHO**

cubic, $Im\overline{3}m$, $a = 15.031Å$ [2]

Framework density: 14.1 T/1000Å^3

Channels: <100> **8** 3.6 x 3.6*** | <100> **8** 3.6 x 3.6***

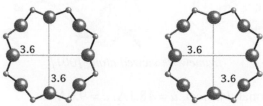

8-ring viewed along <100>

References:
(1) Robson, H.E., Shoemaker, D.P., Ogilvie, R.A. and Manor, P.C. *Adv. Chem. Ser.*, **121**, 106–115 (1973)
(2) McCusker, L.B. and Baerlocher, Ch. *Proc. 6th Int. Zeolite Conf.*, pp. 812–822 (1984)
(3) Gier, T.E. and Stucky, G.D. *Nature*, **349**, 508–510 (1991)
(4) Harvey, G. and Meier, W.M. *Stud. Surf. Sci. Catal.*, **49**, 411–420 (1989)
(5) Feng, P., Bu, X. and Stucky, G.D. *Microporous Mesoporous Mat.*, **23**, 315–322 (1998)
(6) Johnson, G.M., Tripathi, A. and Parise, J.B. *Microporous Mesoporous Mat.*, **28**, 139–154 (1999)
(7) Newsam, J.M., Vaughan, D.E.W. and Strohmaier, K.G. *J. Phys. Chem.*, **99**, 9924–9932 (1995)
(8) Breck, D.W. and Skeels, G.W. *U.S. Patent 4,503,023* (1985)
(9) Rouse, R.C., Peacor, D.R., Dunn, P.J., Campbell, T.J., Roberts, W.L., Wicks, F.J. and Newbury, D. *N. Jb. Miner. Mh.*, 433–440 (1987)
(10) Rouse, R.C., Peacor, D.R. and Merlino, S. *Am. Mineral.*, **74**, 1195–1202 (1989)

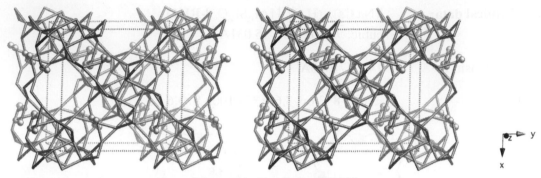

framework viewed along [001]

Idealized cell data: tetragonal, *I4/mcm*, $a = 18.1$Å, $c = 9.0$Å

Coordination sequences and vertex symbols:

T_1 (16,*m*)	4	7	17	31	49	76	98	125	170	208	$3 \cdot 4 \cdot 10_4 \cdot * \cdot 10_4 \cdot *$
T_2 (16,*m*)	4	10	20	31	47	78	109	127	162	212	$4 \cdot 4 \cdot 6 \cdot 6 \cdot 6 \cdot 6$
T_3 (16,2)	4	10	16	31	56	67	94	146	164	188	$4 \cdot 4 \cdot 6_2 \cdot 10_4 \cdot 10_2 \cdot 10_2$
T_4 (8,*m*. 2*m*)	2	4	10	24	38	58	91	110	138	194	3

Secondary building units: see *Compendium*

Composite building units:

  vsv   *lau*

Materials with this framework type:
   *Roggianite[1]

Roggianite   **Type Material Data**   **-RON**

Crystal chemical data: |Ca$_{16}$ (H$_2$O)$_{19}$| [Al$_{16}$Be$_8$Si$_{32}$ O$_{104}$ (OH)$_{16}$]- **-RON**
tetragonal, *I4/mcm*, *a* = 18.33Å, *c* = 9.16Å [1]

Framework density: 18.2 T/1000Å^3

Channels: [001] **12** 4.3 x 4.3*

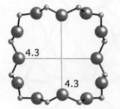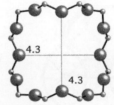

12-ring viewed along [001]

References:
(1)   Giuseppetti, G., Mazzi, F., Tadini, C. and Galli, E. *N. Jb. Miner. Mh.*, **7**, 307–314 (1991)

RRO

Framework Type Data

P2/c

framework viewed along [100]

Idealized cell data: monoclinic, $P2/c$, $a = 7.4$Å, $b = 8.6$Å, $c = 17.2$Å, $\beta = 113.7°$

Coordination sequences and vertex symbols:

T_1 (4,1)	4	11	23	40	56	83	130	163	182	226	300	371		$4·8·5·8·5·10_4$
T_2 (4,1)	4	10	19	37	59	86	111	149	201	241	281	340		$4·5·4·8·5·5$
T_3 (4,1)	4	11	21	36	62	90	111	148	203	252	285	331		$4·5·5·5·5·10_3$
T_4 (4,1)	4	10	20	34	62	86	108	150	201	242	285	329		$4·5·4·5·5·8$
T_5 (2,2)	4	12	18	34	62	90	112	136	198	264	286	314		$5·5·5_2·5_2·10·10_3$

Secondary building units: 4-4 = 1

Composite building units:
bre

Materials with this framework type:
*RUB-41[1]

RUB-41   **Type Material Data**   **RRO**

Crystal chemical data: [Si$_{18}$O$_{36}$]-**RRO**
                                  monoclinic, *P*2/*c*

$a = 7.345$ Å, $b = 8.724$ Å, $c = 17.152$ Å, $\beta = 114.2°$ [1]

Framework density: 18 T/1000Å^3

Channels: [100] **10** 4.0 x 6.5* ↔ [001] **8** 2.7 x 5.0*

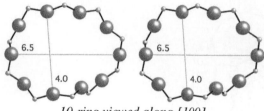
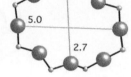
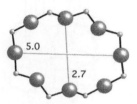

10-ring viewed along [100]   *8-ring viewed along [001]*

References:
(1) Wang, Y., Marler, B., Gies, H. and Müller, U. *Chem. Mater.*, **17**, 43–49 (2005)

RSN

Framework Type Data

Cmmm

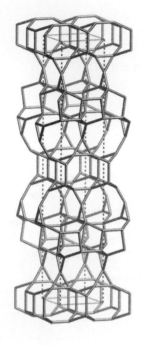
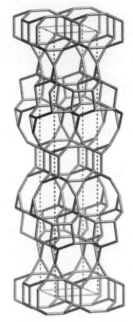

framework viewed along [001]

Idealized cell data: orthorhombic, *Cmmm*, a = 7.2Å, b = 41.9Å, c = 7.2Å

Coordination sequences and vertex symbols:

T_1 (8,..m)	4	9	19	39	59	78	117	155	189	235	288	330	$3 \cdot 4 \cdot 8_3 \cdot 9_4 \cdot 8_3 \cdot 9_4$
T_2 (8,m..)	4	10	21	37	58	91	117	144	194	241	287	343	$4 \cdot 4 \cdot 6_2 \cdot 8 \cdot 6_2 \cdot 8$
T_3 (8,m..)	4	9	21	42	57	82	119	151	188	239	277	345	$3 \cdot 4 \cdot 8_2 \cdot 9_4 \cdot 8_2 \cdot 9_4$
T_4 (8,..m)	4	11	21	40	61	89	116	145	191	239	294	339	$4 \cdot 5_2 \cdot 5 \cdot 8 \cdot 5 \cdot 8$
T_5 (4,$m2m$)	4	8	20	44	55	80	118	152	204	228	272	354	$3 \cdot 3 \cdot 9_4 \cdot 9_4 \cdot 9_4 \cdot 9_4$

Secondary building units: see *Compendium*

Composite building units:

lov   *vsv*

Materials with this framework type:
    *RUB-17[1]

RUB-17 **Type Material Data** **RSN**

Crystal chemical data: $|K_4Na_{12}(H_2O)_{18}|$ $[Zn_8Si_{28}O_{72}]$-**RSN**

monoclinic, *Cm*, $a = 7.238Å$, $b = 40.56Å$, $c = 7.308Å$, $\beta = 91.8°$ [1]

Stability: Complete dehydration leads to destruction of the framework [1]

Framework density: 16.8 T/1000Å^3

Channels: [100] **9** 3.3 x 4.4* ↔ [001] **9** 3.1 x 4.3* ↔[010] **8** 3.4 x 4.1*

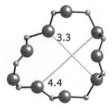 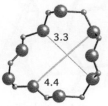 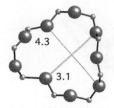 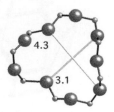

9-ring viewed along [100] *9-ring viewed along [001]*

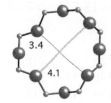 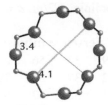

8-ring viewed along [010]

References:
(1) Röhrig, C. and Gies, H. *Angew. Chem., Int. Ed.*, **34**, 63–65 (1995)

RTE

Framework Type Data

C2/m

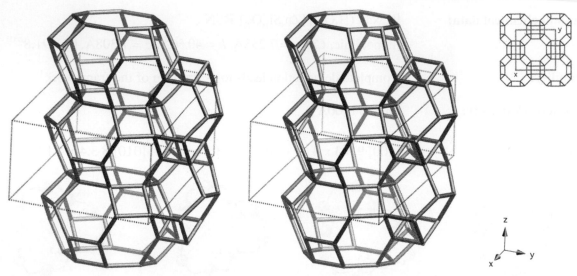

framework viewed normal to [001] (upper right: projection down [001])

Idealized cell data: monoclinic, *C2/m*, $a = 14.1$Å, $b = 13.7$Å, $c = 7.4$Å, $\beta = 102.4°$

Coordination sequences and vertex symbols:

T_1 (8,1)	4	10	19	33	56	81	105	136	175	219	4·5·4·6·5·6
T_2 (8,1)	4	10	22	37	54	79	108	140	176	215	4·4·5·8·6·6
T_3 (8,1)	4	11	21	35	57	80	106	139	176	218	4·5·5·6·6·8

Secondary building units: 6 or 5-1

Composite building units:

rte

Materials with this framework type:
   *RUB-3[1,2]

RUB-3   **Type Material Data**   **RTE**

Crystal chemical data: $|(C_8H_{15}N)_2|$ $[Si_{24}O_{48}]$-**RTE**
                         $C_8H_{15}N$ = exo-2-aminobicyclo[2.2.1.]heptane
                 monoclinic, $C2/m$

                     a = 14.039Å, b = 13.602Å, c = 7.428Å, β = 102.22° [1]

Framework density: 17.3 T/1000Å³

Channels: [001] **8** 3.7 x 4.4*

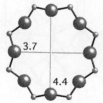 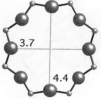

8-ring viewed along [001]

References:
(1)   Marler, B., Grünewald-Luke, A. and Gies, H. *Zeolites*, **15**, 388–399 (1995)
(2)   Marler, B., Grünewald-Lüke, A. and Gies, H. *Microporous Mesoporous Mat.*, **26**, 49–59 (1998)

RTH

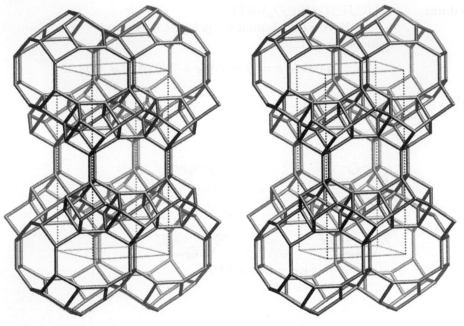

framework viewed along [001]

Idealized cell data: monoclinic, *C2/m*, *a* = 9.8Å, *b* = 20.5Å, *c* = 10.0Å, β = 96.9°

Coordination sequences and vertex symbols:

T$_1$ (8,1)	4	11	21	34	53	78	108	137	165	207		4·6·5·6·5·8
T$_2$ (8,1)	4	10	21	36	54	75	100	136	181	214		4·4·5·8·5·8
T$_3$ (8,1)	4	10	19	31	50	82	106	130	168	203		4·5·4·6·5·5
T$_4$ (8,1)	4	10	18	31	55	77	103	134	165	214		4·5·4·8·5·5

Secondary building units: 4

Composite building units:
  rth

Materials with this framework type:
  *RUB-13[1]
  SSZ-36 (**ITE-RTH** structural intermediate)[2]
  SSZ-50 ([Al-Si-O]-**RTH**)[2]

RUB-13   **Type Material Data**   **RTH**

Crystal chemical data: | $(C_{10}H_{21}N)_2$| $[B_2Si_{30}O_{64}]$-**RTH**
                                    $C_{10}H_{21}N$ = pentamethylpiperidinium
                                    monoclinic, $C2/m$

                                    $a = 9.659Å, b = 20.461Å, c = 9.831Å, β = 96.58°^{(1)}$

Framework density: 16.6 T/1000Å^3

Channels: [100] **8** 3.8 x 4.1* ↔ [001] **8** 2.5 x 5.6*

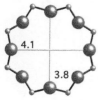 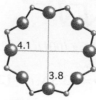 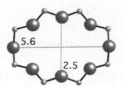 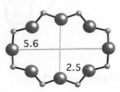

8-ring viewed along [100]   *8-ring viewed along [001]*

References:
(1) Vortmann, S., Marler, B., Gies, H. and Daniels, P. *Microporous Materials*, **4**, 111–121 (1995)
(2) Wagner, P., Nakagawa, Y., Lee, G.S., Davis, M.E., Elomari, S., Medrud, R.C. and Zones, S.I. *J. Am. Chem. Soc.*, **122**, 263–273 (2000)
(3) Lee, G.S. and Zones, S.I. *J. Solid State Chem.*, **167**, 289–298 (2002)

RUT

Framework Type Data

C2/m

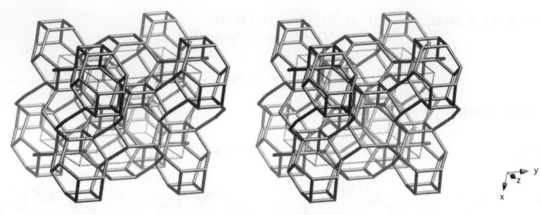

framework viewed along [001]

Idealized cell data: monoclinic, *C2/m*, *a* = 13.2Å, *b* = 13.3Å, *c* = 12.5Å, β = 114.8°

Coordination sequences and vertex symbols:

T$_1$ (8,1)	4	10	21	37	58	87	116	146	185	232	4·4·5·6·5·8
T$_2$ (8,1)	4	10	21	38	60	84	113	148	192	232	4·5·4·6·5·6
T$_3$ (8,1)	4	11	23	38	58	86	114	148	189	234	4·5·5·6·6·6
T$_4$ (8,1)	4	11	21	37	62	85	114	148	185	232	4·5·5·5·6·8
T$_5$ (4,2)	4	12	22	40	58	82	116	154	186	232	5·5·6·6·6·6

Secondary building units: 6

Composite building units:

rte

Materials with this framework type:

*RUB-10[1]

|TMA-|[Si-O]-**RUT**[2]

Al-Nu-1[3]

B-Nu-1[3]

Fe-Nu-1[3]

Ga-Nu-1[3]

Nu-1[4]

RUB-10   **Type Material Data**   **RUT**

Crystal chemical data: $|(C_4H_{12}N)_4| [B_4Si_{32}O_{72}]$-**RUT**

                                  $C_4H_{12}N$ = tetramethylammonium

                            monoclinic, $P2_1/a$

                                  $a = 13.112$Å, $b = 12.903$Å, $c = 12.407$Å, $\beta = 113.50°$ [1]

Framework density: 18.7 T/1000Å^3

Channels: apertures formed by 6-rings only

References:

(1)   Gies, H. and Rius, J. *Z. Kristallogr.*, **210**, 475–480 (1995)

(2)   Broach, R.W., McGuire, N.K., Chao, C.C. and Kirchner, R.M. *J. Phys. Chem. Solids*, **56**, 1363–1368 (1995)

(3)   Bellussi, G., Millini, R., Carati, A., Maddinelli, G. and Gervasini, A. *Zeolites*, **10**, 642–649 (1990)

(4)   Whittam, T.V. and Youll, B. *U.S. Patent 4,060,590* (1977)

RWR

Framework Type Data

I4₁/amd

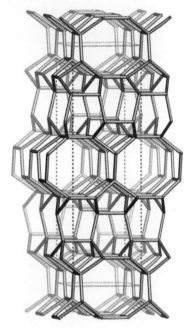 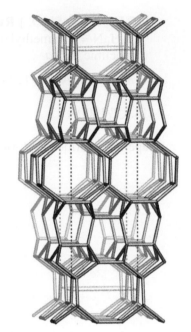

framework viewed along [010]

Idealized cell data: tetragonal, *I4₁/amd* (origin choice 2), *a* = 7.8Å, *c* = 27.3Å

Coordination sequences and vertex symbols:

T₁ (16,.m.)	4	12	24	42	68	97	133	180	221	277	334	394		5·6·5·6·5·6₂
T₂ (16,..2)	4	11	23	42	66	94	133	173	218	280	328	393		4·5₂·6₂·6₂·8·8

Secondary building units: 6-2

Composite building units:

  mor

Materials with this framework type:
  *RUB-24[1,2]

RUB-24   **Type Material Data**   **RWR**

Crystal chemical data: [Si$_{32}$O$_{64}$]-**RWR**
tetragonal, *I4$_1$/amd*, *a* = 7.6677Å, *c* = 27.0625Å [2]

Stability: stable at 900°C [0]

Framework density: 20.1 T/1000Å^3

Channels: [100] **8** 2.8 x 5.0* ‖ [010] **8** 2.8 x 5.0*
(nonintersecting 1-d 8-ring channels)

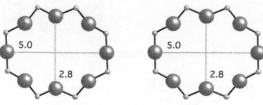

8-ring viewed along <100>

References:
(1) Marler, B., Ströter, N. and Gies, H. Recent Research Reports, 14th IZC, Cape Town, South Africa, 15–16 (2004)
(2) Marler, B., Ströter, N. and Gies, H. Microporous Mesoporous Mat., **83**, 201–211 (2005)

RWY Framework Type Data *Im3̄m*

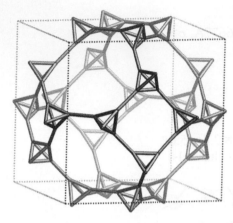 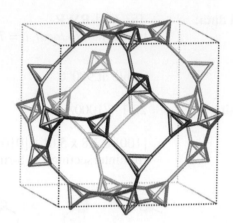

framework viewed along [100]

Idealized cell data: cubic, $Im\bar{3}m$, $a = 18.5$Å

Coordination sequences and vertex symbols:

| T_1 (48,*m*) | 4 | 6 | 12 | 17 | 28 | 38 | 52 | 64 | 84 | 104 | 124 | 143 | | 3·8·3·12·3·12 |

Secondary building units: 8 or 3*1

Materials with this framework type:
    *UCR-20[1]

UCR-20 **Type Material Data** **RWY**

Crystal chemical data: $|(C_6H_{18}N_4)_{16}|$ $[Ga_{32}Ge_{16}S_{96}]$-**RWY**
$C_6H_{18}N_4$ = TAEA = tris(2-aminoethyl)amine
cubic, $I\overline{4}3m$, $a = 20.9352$Å [1]

Stability: Stable when heated in air at 300ºC for 1h (ca 20% weight loss) or at 380ºC in argon [1]

Framework density: 5.2 T/1000Å^3

Channels: <111> **12** 6.9 x 6.9***

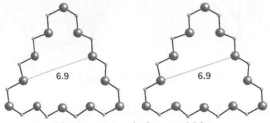

12-ring viewed along <111>

References:
(1) Zheng, N., Bu, X., Wang, B. and Feng, P. *Science*, **298**, 2366–2369 (2002)

SAO

Framework Type Data

$I\overline{4}m2$

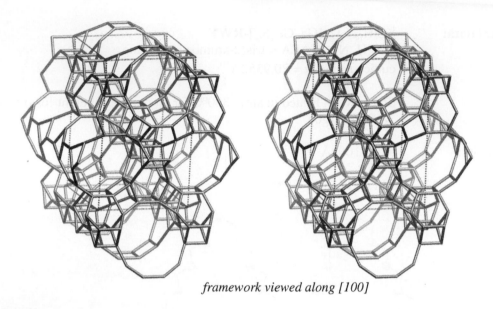

framework viewed along [100]

Idealized cell data: tetragonal, $I\overline{4}m2$, $a = 13.4$Å, $c = 21.9$Å

Coordination sequences and vertex symbols:

T_1 (16,1)	4	9	16	25	39	61	84	102	124	158	$4 \cdot 4 \cdot 4 \cdot 6_2 \cdot 6_3 \cdot 12_4$
T_2 (16,1)	4	9	17	27	40	61	85	106	132	167	$4 \cdot 4 \cdot 4 \cdot 12_5 \cdot 6 \cdot 6_3$
T_3 (16,1)	4	9	16	25	39	58	79	104	130	158	$4 \cdot 6 \cdot 4 \cdot 6 \cdot 4 \cdot 12_6$
T_4 (8,2)	4	10	16	25	42	61	82	108	132	156	$4 \cdot 4 \cdot 6 \cdot 6 \cdot 6_2 \cdot 12_5$

Secondary building units: 4

Composite building units:

dsc	*sti*	*lau*	*aww*
double sawtooth chain			

Materials with this framework type:
  *STA-1[1]

STA-1 **Type Material Data** **SAO**

Crystal chemical data: $|(C_{21}H_{40}N_2)_{2.6}(H_2O)_6|[Mg_5Al_{23}P_{28}O_{112}]$-**SAO**

$C_{21}H_{40}N_2 = C_7H_{13}N - (CH_2)_7 - C_7H_{13}N$

$C_7H_{13}N$ = quinuclidine

tetragonal, $P\bar{4}n2$, $a = 13.810$Å, $c = 21.969$Å [1]

Framework density: 13.4 T/1000Å^3

Channels: <100> **12** 6.5 x 7.2** $\leftrightarrow$ [001] **12** 7.0 x 7.0*

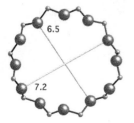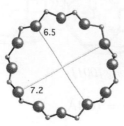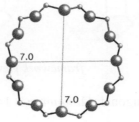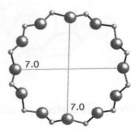

12-ring viewed along <100>   *12-ring viewed along [001]*

References:
(1) Noble, G.W., Wright, P.A., Lightfoot, P., Morris, R.E., Hudson, K.J., Kvick, A. and Graafsma, H. *Angew. Chem., Int. Ed.*, **36**, 81–83 (1997)

SAS

Framework Type Data

I4/mmm

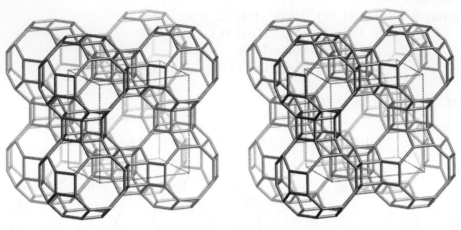

framework viewed along [001]

Idealized cell data: tetragonal, *I4/mmm*, $a = 14.3$Å, $c = 10.4$Å

Coordination sequences and vertex symbols:

T$_1$ (16,*m*)	4	10	19	30	45	65	90	118	145	175	4·6·4·6·6·8
T$_2$ (16,2)	4	9	17	30	48	68	87	109	142	184	4·4·4·6·6·6

Secondary building units: 6-2 or 4

Composite building units:
   d6r

Materials with this framework type:
   *STA-6[1]
   SSZ-73[2]

STA-6 Type Material Data **SAS**

Crystal chemical data: $|(C_{14}H_{34}N_4)_{1.5} (H_2O)_{2.5}|$ $[Mg_3Al_{13}P_{16}O_{64}]$-**SAS**
$C_{14}H_{32}N_4$ = 1,4,8,11-tetramethyl-1,4,8,11-tetraazatetradecane
tetragonal, $P4/mnc$, a = 14.282Å, c = 10.249Å [1]

Framework density: 15.3 T/1000Å^3

Channels: [001] **8** 4.2 x 4.2*

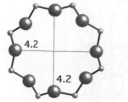 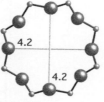

8-ring viewed along [001]

References:
(1) Patinec, V., Wright, P.A., Lightfoot, P., Aitken, R.A. and Cox, P.A. *J. Chem. Soc., Dalton Trans.*, 3909–3911 (1999)
(2) Zones, S., Burton, A. and Ong, K. *U.S. Patent 7,138,099* (2006)

SAT

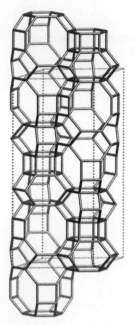
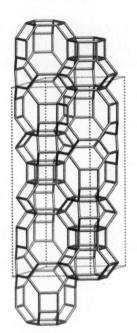
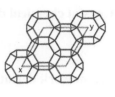

framework viewed normal to [001] (upper right: projection down [001])

Idealized cell data: trigonal, $R\bar{3}m$, $a = 12.9$Å, $c = 30.6$Å

Coordination sequences and vertex symbols:

T_1 (36,1)	4	10	20	33	50	71	95	124	158	197	4·6·4·8·6·6
T_2 (36,1)	4	9	17	30	50	75	100	126	157	194	4·4·4·6·6·8

Secondary building units: 6 or 4

Framework description: ABBCBCCACAAB sequence of 6-rings

Composite building units:

    d6r   *can*

Materials with this framework type:
    *STA-2[1]

STA-2 | **Type Material Data** | **SAT**

Crystal chemical data: $|(C_{18}H_{34}N_2)_3 (H_2O)_{22.5}| [Mg_{5.4}Al_{30.6}P_{36}O_{144}]$-**SAT**
$C_{18}H_{34}N_2 = C_7H_{13}N - (CH_2)_4 - C_7H_{13}N$
$C_7H_{13}N$ = quinuclidine
trigonal, $R\overline{3}$, $a = 12.726$ Å, $c = 30.939$Å [1]

Framework density: 16.6 T/1000Å^3

Channels: $\perp$ [001] **8** 3.0 x 5.5***

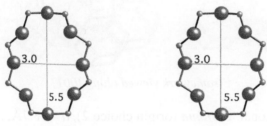

8-ring viewed normal to [001]

References:
(1)   Noble, G.W., Wright, P.A. and Kvick, A. *J. Chem. Soc., Dalton Trans.*, 4485–4490 (1997)

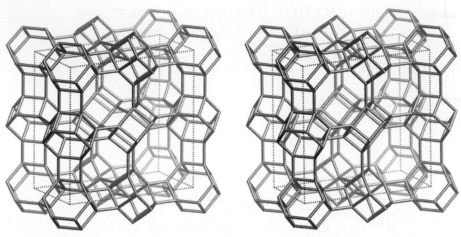

framework viewed along [001]

Idealized cell data: tetragonal, *P4/nmm* (origin choice 2), $a = 18.7$Å, $c = 9.4$Å

Coordination sequences and vertex symbols:

T_1 (16,1)	4	9	17	29	45	65	88	113	143	179	4·4·4·8·6·8
T_2 (16,1)	4	9	17	29	45	65	88	114	144	177	4·4·4·8·6·8
T_3 (16,1)	4	9	17	29	45	63	84	112	144	177	4·4·4·8·6·8

Secondary building units: 6-6 or 6 or 4-2 or 4

Composite building units:
    d6r

Materials with this framework type:
    *Mg-STA-7[1]
    Co-STA-7[1]
    Zn-STA-7[1]

Mg-STA-7   **Type Material Data**   **SAV**

Crystal chemical data: $|(C_{18}H_{42}N_6)_{1.96}(H_2O)_7|\,[Mg_{4.8}Al_{19.2}P_{24}O_{96}]$-**SAV**

$C_{18}H_{42}N_6 = $ 1,4,7,10,13,16-hexamethyl-1,4,7,10,13,16-hexaazacyclooctadecane

tetragonal, $P4/n$, $a = 18.773$Å, $c = 9.454$Å $^{(1)}$

Framework density: 14.4 T/1000Å^3

Channels: <100> **8** 3.8 x 3.8** $\leftrightarrow$ [001] **8** 3.9 x 3.9*

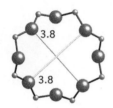 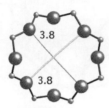 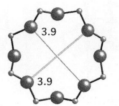 

8-ring viewed along <100>   *8-ring viewed along [001]*

References:

(1) Wright, P.A., Maple, M.J., Slawin, A.M.Z., Patinec, V., Aitken, R.A., Welsh, S. and Cox, P.A. *J. Chem. Soc., Dalton Trans.*, 1243–1248 (2000)

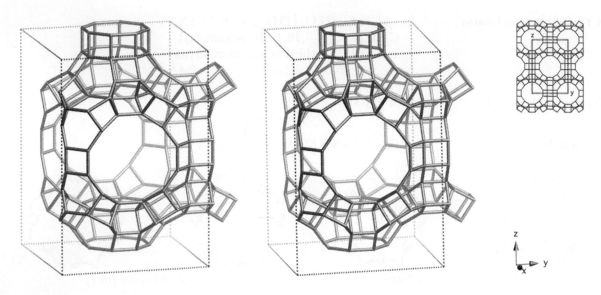

framework viewed along [100] (upper right: projection down [100])

Idealized cell data: tetragonal, *I4/mmm*, $a = 18.5$Å, $c = 27.1$Å

Coordination sequences and vertex symbols:

T_1 (32,1)	4	9	17	27	38	55	78	102	129	157		$4 \cdot 6 \cdot 4 \cdot 8 \cdot 4 \cdot 8_7$
T_2 (32,1)	4	9	17	28	41	57	77	101	130	162		$4 \cdot 4 \cdot 4 \cdot 6 \cdot 8 \cdot 12$
T_3 (32,1)	4	9	17	27	39	56	77	100	126	157		$4 \cdot 4 \cdot 4 \cdot 8 \cdot 6 \cdot 6_2$
T_4 (32,1)	4	9	17	27	40	59	79	99	126	158		$4 \cdot 4 \cdot 4 \cdot 8_2 \cdot 6_2 \cdot 8_4$

Secondary building units: 8 or 4

Composite building units:

    sti   *d8r*   *atn*

Materials with this framework type:
    *UCSB-8Co[(1)]
    UCSB-8Mg[(1)]
    UCSB-8Mn[(1)]
    UCSB-8Zn[(1)]

UCSB-8Co   Type Material Data   **SBE**

Crystal chemical data: $|(C_9H_{24}N_2)_{16}|$ $[Al_{32}Co_{32}P_{64}O_{256}]$-**SBE**
                                $C_9H_{22}N_2 = $ 1,9 diaminononane
                                tetragonal, $P4/nnc$, $a = 19.065$Å, $c = 27.594$ Å [1]

Framework density: 12.8 T/1000Å^3

Channels: <100> **12** 7.2 x 7.4** ↔ [001] **8** 4.0 x 4.0*

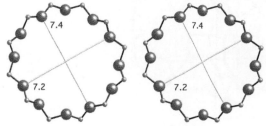 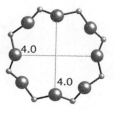

12-ring viewed along <100>   *8-ring viewed along [001]*

second 8-ring along [001]

References:
(1) Bu, X., Feng, P. and Stucky, G.D. *Science*, **278**, 2080–2085 (1997)

SBS

Framework Type Data

P6₃/mmc

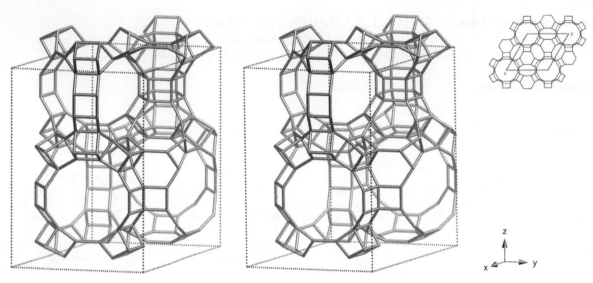

framework viewed normal to [001] (upper right: projection down [001])

Idealized cell data: hexagonal, $P6_3/mmc$, $a = 17.2$Å, $c = 27.3$Å

Coordination sequences and vertex symbols:

T_1 (24,1)	4	9	17	28	41	56	75	100	127	157	195	233	$4 \cdot 4 \cdot 4 \cdot 12_6 \cdot 6_2 \cdot 8_4$
T_2 (24,1)	4	9	17	27	39	55	75	100	127	156	192	230	$4 \cdot 4 \cdot 4 \cdot 8 \cdot 6 \cdot 6_2$
T_3 (24,1)	4	9	16	25	38	58	84	111	135	157	182	215	$4 \cdot 4 \cdot 4 \cdot 6 \cdot 6 \cdot 12$
T_4 (24,1)	4	9	16	24	35	53	77	104	130	153	178	213	$4 \cdot 6 \cdot 4 \cdot 6 \cdot 4 \cdot 8_7$

Secondary building units: 8 or 4

Composite building units:

sti   d6r   can

Materials with this framework type:

*UCSB-6GaCo[1]   UCSB-6Mg[1]
UCSB-6Co[1]   UCSB-6Mn[1]
UCSB-6GaMg[1]   UCSB-6Zn[1]
UCSB-6GaZn[1]

UCSB-6GaCo   **Type Material Data**   **SBS**

Crystal chemical data: $|(C_9H_{24}N_2)_{12}|$ $[Ga_{24}Co_{24}P_{48}O_{192}]$-**SBS**
$C_9H_{22}N_2$ = 1,9-diaminononane
trigonal, $P\bar{3}1c$, $a = 17.836$Å, $c = 27.182$Å [1]

Framework density: 12.8 T/1000Å^3

Channels: [001] **12** 6.8 x 6.8* $\leftrightarrow$ $\perp$ [001] **12** 6.9 x 7.0**

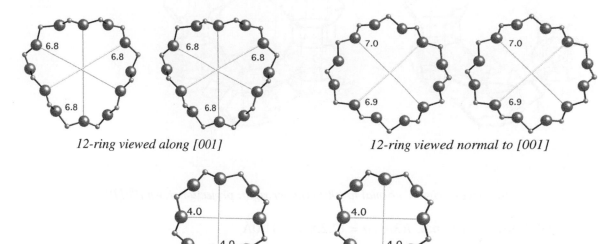

12-ring viewed along [001]   *12-ring viewed normal to [001]*

8-ring along [001]

References:
(1)   Bu, X., Feng, P. and Stucky, G.D. *Science*, **278**, 2080–2085 (1997)

SBT

Framework Type Data

$R\bar{3}m$

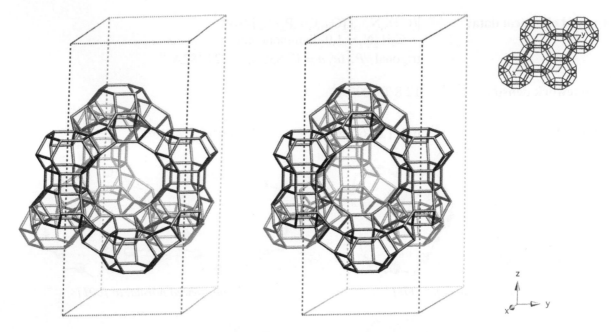

framework viewed normal to [001] (upper right: projection down [001])

Idealized cell data: trigonal, $R\bar{3}m$, a = 17.2Å, c = 41.0Å

Coordination sequences and vertex symbols:

T_1 (36,1)	4	9	16	25	38	58	84	111	135	157	182	215		$4 \cdot 4 \cdot 4 \cdot 6 \cdot 6 \cdot 12$
T_2 (36,1)	4	9	17	28	41	56	75	100	127	157	195	232		$4 \cdot 4 \cdot 4 \cdot 12_6 \cdot 6_2 \cdot 8_4$
T_3 (36,1)	4	9	17	27	39	55	75	100	127	156	192	228		$4 \cdot 4 \cdot 4 \cdot 8 \cdot 6 \cdot 6_2$
T_4 (36,1)	4	9	16	24	35	53	77	104	130	153	178	213		$4 \cdot 6 \cdot 4 \cdot 6 \cdot 4 \cdot 8_7$

Secondary building units: 4

Composite building units:

     sti   *d6r*   *can*

Materials with this framework type:
   *UCSB-10GaZn[1]
   UCSB-10Co[1]
   UCSB-10Mg[1]
   UCSB-10Zn[1]

UCSB-10GaZn   Type Material Data   SBT

Crystal chemical data: |(C$_{10}$H$_{26}$N$_2$O$_3$)$_{18}$| [Ga$_{36}$Zn$_{36}$P$_{72}$O$_{288}$]-**SBT**
C$_{10}$H$_{26}$N$_2$O$_3$ = 4,7,10-trioxa-1, 13-tridecanediammonium
trigonal, $R\bar{3}$, , a = 18.080Å, c = 41.951Å [1]

Framework density: 12.1 T/1000Å^3

Channels: [001] **12** 6.4 x 7.4 * ↔ ⊥ [001] **12** 7.3 x 7.8**

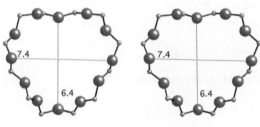

12-ring viewed along [001]

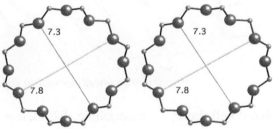

12-ring viewed normal to [001]

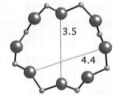

8-ring viewed along [102]

References:
(1) Bu, X., Feng, P. and Stucky, G.D. *Science*, **278**, 2080–2085 (1997)

SFE

Framework Type Data

P2₁/m

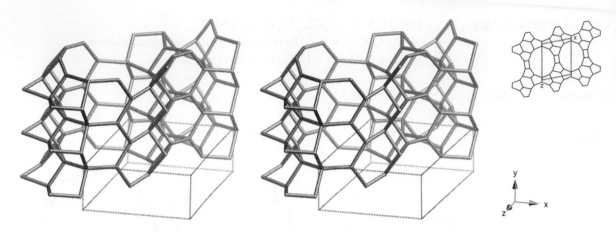

framework viewed normal to [010] (upper right: projection down [010])

Idealized cell data: monoclinic, $P2_1/m$, $a = 11.5$Å, $b = 5.3$Å, $c = 14.0$Å, $\beta = 101°$

Coordination sequences and vertex symbols:

T₁ (2,m)	4	12	20	37	62	82	114	142	192	238	278	316	5·5·5·5·6₂·12₂
T₂ (2,m)	4	10	19	35	58	86	108	144	183	233	287	322	4·5·4·5·6·12₂
T₃ (2,m)	4	10	20	35	57	83	116	141	181	230	284	342	4·5·4·5·12₂·*
T₄ (2,m)	4	12	22	36	56	86	114	150	179	232	283	337	5·5·5·5·6₂·*
T₅ (2,m)	4	12	22	37	55	83	120	149	177	227	288	347	5₂·6₂·6·6₂·6·6₂
T₆ (2,m)	4	12	24	37	54	80	117	158	184	219	281	338	5₂·6₂·6·6₂·6·6₂
T₇ (2,m)	4	12	23	41	58	78	111	154	198	235	261	318	5·5·5·5·6·12₂

Secondary building units: see *Compendium*

Composite building units:

dzc	jbw	mtt	bik	ton
double zigzag chain				

Materials with this framework type:
  *SSZ-48[1]

SSZ-48   **Type Material Data**   **SFE**

Crystal chemical data: [Si$_{14}$O$_{28}$]-**SFE**
                                monoclinic, $P2_1$

a = 11.153Å, b = 5.002Å, c = 13.667Å, β = 100.63° [1]

Framework density: 18.7 T/1000Å^3

Channels: [010] **12** 5.4 x 7.6*

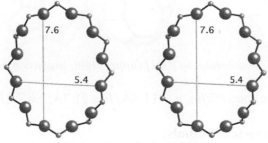

12-ring viewed along [010]

References:
(1) Wagner, P., Terasaki, O., Ritsch, S., Nery, J.G., Zones, S.I., Davis, M.E. and Hiraga, K. *J. Phys. Chem. B*, **103**, 8245–8250 (1999)

SFF

Framework Type Data

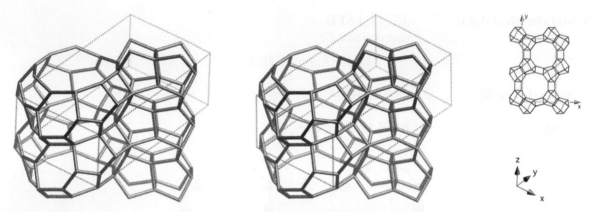

framework viewed normal to [001] (upper right: projection down [001])

Idealized cell data: monoclinic, $P2_1/m$, $a = 11.5$Å, $b = 21.7$Å, $c = 7.2$Å, $\beta = 93.2°$

Coordination sequences and vertex symbols:

T_1 (4,1)	4	12	20	34	56	87	115	143	176	224	5·5$_2$·5·6·5·6
T_2 (4,1)	4	12	20	34	56	88	115	142	177	225	5·5$_2$·5·6·5·6
T_3 (4,1)	4	11	22	38	57	80	111	148	189	228	4·5·5·6·5·10
T_4 (4,1)	4	11	22	39	54	84	110	145	189	234	4·5·5·6·5·10
T_5 (4,1)	4	11	23	37	57	82	113	150	184	228	4·5·5·6·5·10
T_6 (4,1)	4	11	20	31	58	86	115	142	174	225	4·6$_2$·5·5·5·5
T_7 (4,1)	4	11	19	36	55	82	113	148	181	220	4·6·5·5·5·5
T_8 (4,1)	4	11	23	36	59	79	114	147	183	229	4·5·5·6·5·10

Secondary building units: 5-3

Composite building units:

cas   *stf*

Materials with this framework type:
 *SSZ-44[1]
 STF-SFF structural intermediates[2]

SSZ-44   **Type Material Data**   # SFF

Crystal chemical data: $[Si_{32}O_{64}]$-**SFF**

monoclinic, $P2_1/m$

$a = 11.485$ Å, $b = 21.946$Å, $c = 7.388$Å, $\beta = 94.70°$ [1]

Framework density: 17.2 T/1000Å^3

Channels: [001] **10** 5.4 x 5.7*

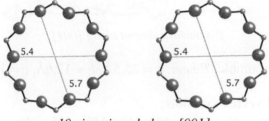

10-ring viewed along [001]

References:
(1) Wagner, P., Zones, S.I., Davis, M.E. and Medrud, R.C. *Angew. Chem., Int. Ed.*, **38**, 1269–1272 (1999)
(2) Villaescusa, L.A., Zhou, W., Morris, R.E. and Barrett, P.A. *J. Mater. Chem.*, **14**, 1982–1987 (2004)

SFG

Framework Type Data

Pmma

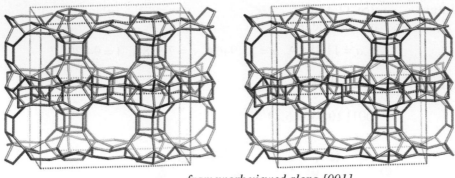

framework viewed along [001]

Idealized cell data: orthorhombic, *Pmma*, $a = 25.5$Å, $b = 12.6$Å, $c = 13.1$Å

Coordination sequences and vertex symbols:

T$_1$ (8,1)	4	12	24	38	61	83	117	159	205	246	293	359		5·6·5·7·6·6$_2$
T$_2$ (8,1)	4	11	22	38	55	79	114	158	201	242	298	355		4·7$_2$·5·6·5·6$_2$
T$_3$ (8,1)	4	12	22	35	61	86	121	155	199	248	306	360		5·6·5·6·6·6$_2$
T$_4$ (8,1)	4	10	20	37	62	90	120	155	193	241	304	374		4·6$_2$·4·6$_2$·5·10
T$_5$ (4,.m.)	4	12	24	36	53	85	120	150	192	241	294	371		5·5$_2$·6·6$_2$·6·6$_2$
T$_6$ (4,.m.)	4	12	22	35	52	88	118	148	182	244	307	363		5·5$_2$·6$_2$·6$_2$·6$_2$·6$_2$
T$_7$ (8,1)	4	11	21	37	56	85	122	161	192	240	305	364		4·6·5·6$_2$·5·6$_2$
T$_8$ (8,1)	4	10	21	40	63	88	118	156	200	247	299	361		4·6·4·6$_2$·5·10$_2$
T$_9$ (4,m..)	4	12	24	36	56	80	116	156	193	235	303	367		5·6·5·6·6$_2$·7$_2$
T$_{10}$ (2,mm2)	4	12	22	36	52	78	122	148	180	236	306	364		5·5$_2$·6$_2$·6$_2$·6$_2$·6$_2$
T$_{11}$ (8,1)	4	10	20	36	58	83	112	152	206	250	288	344		4·6·4·6$_2$·5·6
T$_{12}$ (4,m..)	4	10	20	38	61	86	117	152	196	246	304	360		4·6$_2$·4·6$_2$·5·10$_2$

Secondary building units: see *Compendium*

Composite building units:
   mel

Materials with this framework type:
   *SSZ-58[1]

SSZ-58   **Type Material Data**   **SFG**

Crystal chemical data: [$B_{2.25}Si_{71.75}O_{148}$]-**SFG**
orthorhombic, *Pmma*, a = 25.1118Å, b = 12.4976Å, c = 12.8598Å [1]

Framework density: 18.3 T/1000Å^3

Channels: [001] **10** 5.2 x 5.7* ↔ [100] **10** 4.8 x 5.7*

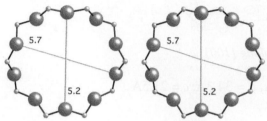

10-ring viewed along [001]

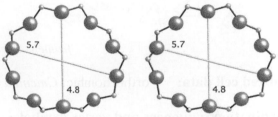

10-ring viewed along [100]

References:
(1) Burton, A., Elomari, S., Medrud, R.C., Chan, I.Y., Chen, C.-Y., Bull, L.M. and Vittoratos, E.S. *J. Am. Chem. Soc.*, **125**, 1633–1642 (2003)

SFH

Cmcm

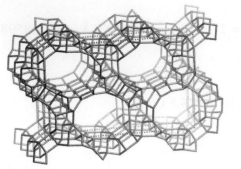
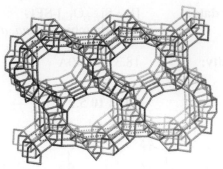

framework viewed along [100]

Idealized cell data: orthorhombic, *Cmcm*, $a = 5.3$Å, $b = 34.3$Å, $c = 21.5$Å

Coordination sequences and vertex symbols:

T_1 (8,*m*..)	4	11	22	35	57	75	107	135	178	221	269	315	$4 \cdot 6_2 \cdot 5 \cdot 6 \cdot 5 \cdot 6$
T_2 (8,*m*..)	4	10	20	31	54	71	106	132	174	212	258	307	$4 \cdot 5 \cdot 4 \cdot 5 \cdot 14_6 \cdot *$
T_3 (8,*m*..)	4	12	22	35	49	80	109	144	178	205	244	321	$5 \cdot 6_2 \cdot 5 \cdot 6_2 \cdot 5_2 \cdot 6$
T_4 (8,*m*..)	4	10	19	34	47	76	100	140	172	213	249	299	$4 \cdot 5 \cdot 4 \cdot 5 \cdot 6 \cdot 14_6$
T_5 (8,*m*..)	4	11	22	38	52	78	103	140	178	220	262	312	$4 \cdot 6_2 \cdot 5 \cdot 6 \cdot 5 \cdot 6$
T_6 (8,*m*..)	4	12	19	35	54	79	103	145	167	216	273	315	$5 \cdot 6 \cdot 5 \cdot 6 \cdot 5_2 \cdot 6$
T_7 (8,*m*..)	4	12	22	34	50	72	108	147	179	208	248	303	$5 \cdot 6_2 \cdot 5 \cdot 6_2 \cdot 6_2 \cdot 14_6$
T_8 (8,*m*..)	4	12	23	36	53	75	108	143	178	212	258	314	$5 \cdot 6 \cdot 5 \cdot 6 \cdot 6 \cdot 6_2$

Secondary building units: 5-3

Composite building units:

dzc	afi	cas	mtw
double zigzag chain			

Materials with this framework type:
*SSZ-53[1]

SSZ-53   **Type Material Data**   **SFH**

Crystal chemical data: [B$_{1.6}$Si$_{62.4}$O$_{128}$]-**SFH**
monoclinic, C2/c

a = 5.0192Å, b = 33.7437Å, c = 21.1653Å, β = 90.485°[(1)]

Framework density: 17.9 T/1000Å^3

Channels: [001] **14** 6.4 x 8.7*

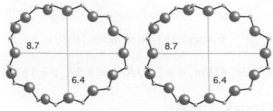

14-ring viewed along [001]

References:
(1) Burton, A., Elomari, S., Chen, C.Y., Medrud, R.C., Chan, I.Y., Bull, L.M., Kibby, C., Harris, T.V., Zones, S.I. and Vittoratos, E.S. *Chem. Eur. Journal*, **9**, 5737–5748 (2003)

SFN

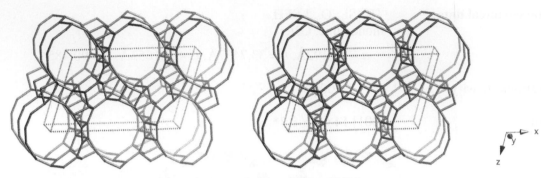

framework viewed along [010]

Idealized cell data: monoclinic, $C2/m$, $a = 25.2$Å, $b = 5.3$Å, $c = 15.0$Å, $\beta = 103.9°$

Coordination sequences and vertex symbols:

T_1 (4,*m*)	4	11	22	35	56	73	103	131	173	217	265	297		$4 \cdot 6_2 \cdot 5 \cdot 6 \cdot 5 \cdot 6$
T_2 (4,*m*)	4	11	22	38	51	76	99	135	176	220	251	296		$4 \cdot 6_2 \cdot 5 \cdot 6 \cdot 5 \cdot 6$
T_3 (4,*m*)	4	12	22	35	51	74	108	146	171	199	242	302		$5 \cdot 6_2 \cdot 5 \cdot 6_2 \cdot 6_2 \cdot 14_6$
T_4 (4,*m*)	4	12	19	34	55	79	101	139	163	209	263	306		$5 \cdot 6 \cdot 5 \cdot 6 \cdot 5_2 \cdot 6$
T_5 (4,*m*)	4	12	23	35	52	73	106	142	176	205	246	295		$5 \cdot 6 \cdot 5 \cdot 6 \cdot 6 \cdot 6_2$
T_6 (4,*m*)	4	12	21	34	49	78	109	138	176	196	238	310		$5 \cdot 6_2 \cdot 5 \cdot 6_2 \cdot 5_2 \cdot 6$
T_7 (4,*m*)	4	10	20	31	53	73	106	130	169	206	252	297		$4 \cdot 5 \cdot 4 \cdot 5 \cdot 14_6 \cdot *$
T_8 (4,*m*)	4	10	19	34	48	76	102	139	166	205	242	299		$4 \cdot 5 \cdot 4 \cdot 5 \cdot 6 \cdot 14_6$

Secondary building units: 5-3

Composite building units:

        dzc   *jbw*   *cas*   *mtw*

double zigzag
chain

Materials with this framework type:
    *SSZ-59[1]

SSZ-59   **Type Material Data**   **SFN**

Crystal chemical data: $[B_{0.35}Si_{15.65}O_{32}]$-**SFN**
                             triclinic, $P\bar{1}$, $a = 5.023$Å, $b = 12.735$Å, $c = 14.722$Å
                             $\alpha = 103.44°$, $\beta = 90.51°$, $\gamma = 100.88°$ [1]

Framework density: 17.8 T/1000Å^3

Channels: [001] **14** 6.2 x 8.5*

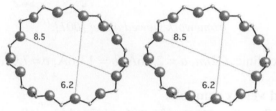

14-ring viewed along [001]

References:
(1) Burton, A., Elomari, S., Chen, C.Y., Medrud, R.C., Chan, I.Y., Bull, L.M., Kibby, C., Harris, T.V., Zones,
     S.I. and Vittoratos, E.S. *Chem. Eur. Journal*, **9**, 5737–5748 (2003)

SFO

Framework Type Data

C2/m

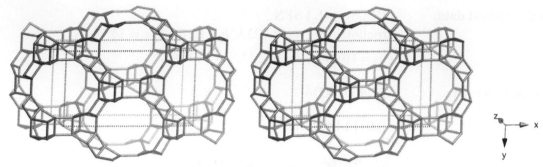

framework viewed along [001]

Idealized cell data: monoclinic, *C2/m*, *a* = 22.6Å, *b* = 13.6Å, *c* = 7.0Å, β = 99°

Coordination sequences and vertex symbols:

T$_1$ (8,1)	4	10	17	28	46	63	86	117	142	168	211	256	4·6·4·6·6·12
T$_2$ (8,1)	4	9	18	30	43	64	90	111	139	178	212	248	4·4·4·8·6$_3$·8
T$_3$ (8,1)	4	9	18	29	42	65	91	111	138	176	210	248	4·4·4·12·6·6$_3$
T$_4$ (8,1)	4	9	16	27	44	65	87	110	138	171	211	257	4·6·4·6$_2$·4·8

Secondary building units: 6-2 or 4-4- or 4

Composite building units:

sti

Materials with this framework type:
   *SSZ-51[1]

SSZ-51   **Type Material Data**   **SFO**

Crystal chemical data: |(C₇N₂H₁₁)₈ (H₂O)₄| [F₈Al₃₂P₃₂O₁₂₈]-**SFO**

$|(C_7N_2H_{11})_8 (H_2O)_4| [F_8Al_{32}P_{32}O_{128}]$-**SFO**
$C_7N_2H_{11}$ = 4-dimethylaminopyridinium
monoclinic, $C2/c$

$a = 21.759$Å, $b = 13.821$Å, $c = 14.224$ Å, $\beta = 98.849°$ [(1)]
(Relationship to unit cell of Framework Type: $a' = a$, $b' = b$, $c' = 2c$)

Framework density: 15.1 T/1000Å^3

Channels: [001] **12** 6.9 x 7.1* ↔ [010] **8** 3.1 x 3.9*

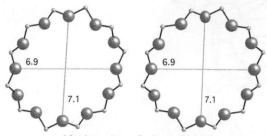

12-ring viewed along [001]

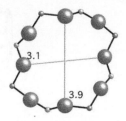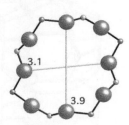

8-ring viewed along [010]

References:
(1) Morris, R.E., Burton, A., Bull, L.M. and Zones, S.I. *Chem. Mater.*, **16**, 2844–2851 (2004)

SGT

Framework Type Data

I4₁/amd

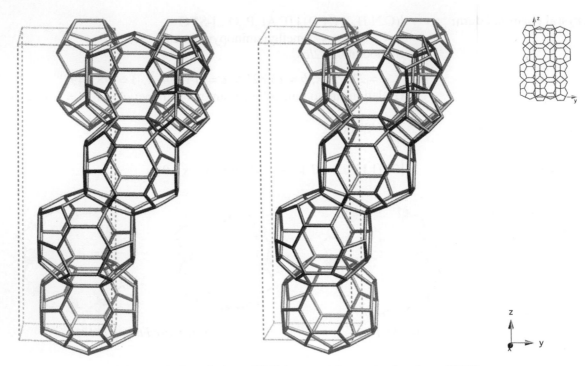

framework viewed along [100] (upper right: projection down [100])

Idealized cell data: tetragonal, *I4₁/amd* (origin choice 2), $a = 10.3$Å, $c = 34.4$Å

Coordination sequences and vertex symbols:

T₁ (16,*m*)	4	11	22	37	62	89	120	155	202	257	4·6·5·5·5·5
T₂ (16,*m*)	4	11	21	37	63	86	121	152	196	258	4·6·5·5·5·5
T₃ (16,2)	4	11	23	38	62	92	113	159	210	244	4·6·5·5·5·5
T₄ (16,*m*)	4	12	24	42	61	87	128	168	205	250	5·6·5·6·5·6

Secondary building units: 5-3

Materials with this framework type:

*Sigma-2[1]

[B-Si-O]-**SGT**[2]

Sigma-2   **Type Material Data**   **SGT**

Crystal chemical data: $|(C_{10}H_{17}N)_4|$ $[Si_{64}O_{128}]$-**SGT**
$C_{10}H_{17}N$ = 1-aminoadamantane
tetragonal, $I4_1/amd$, $a = 10.239$Å, $c = 34.383$Å [1]

Framework density: 17.8 T/1000Å^3

Channels: apertures formed by 6-rings only

References:
(1) McCusker, L.B. *J. Appl. Crystallogr.*, **21**, 305–310 (1988)
(2) Grünewald-Luke, A., Marler, B., Hochgrafe, M. and Gies, H. *J. Mater. Chem.*, **9**, 2529–2536 (1999)

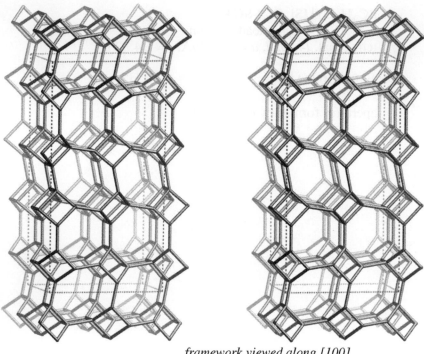

framework viewed along [100]

Idealized cell data: orthorhombic, *Cmcm*, $a = 9.9$Å, $b = 14.1$Å, $c = 28.1$Å

Coordination sequences and vertex symbols:

T_1 (16,1)	4	9	18	32	49	69	93	122	156	191	229	275	$4·4·4·8_2·8·8$
T_2 (16,1)	4	9	18	32	48	67	92	120	150	187	231	275	$4·4·4·8_2·8·8$
T_3 (16,1)	4	9	18	32	48	68	96	126	155	191	234	277	$4·4·4·8_2·8·8$
T_4 (16,1)	4	9	18	32	49	69	93	121	155	193	230	272	$4·4·4·8_2·8·8$

Secondary building units: 4 or 8

Composite building units:

dcc   *gis*

*double
crankshaft chain*

Materials with this framework type:
*SIZ-7[1]

SIZ-7   **Type Material Data**   # SIV

Crystal chemical data: $[Co_{12.8}Al_{19.2}P_{32}O_{128}]$-**SIV**
                              monoclinic, *C2/c*

$a = 10.2959$Å, $b = 14.3715$Å, $c = 28.599$Å, $\beta = 91.094°$ [1]

Framework density: 15.1 T/1000Å^3

Channels: { [100] **8** (3.5 x 3.9+ **8** 3.7 x 3.8 ↔ [110] **8** 3.7 x 3.8

↔ [001] **8** 3.8 x 3.9°} ***

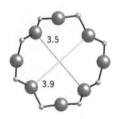 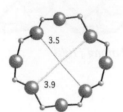 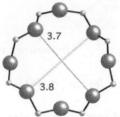 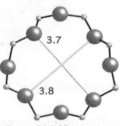

8-ring viewed along [100]   *2nd 8-ring viewed along [100]*

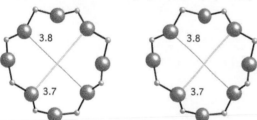  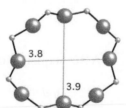 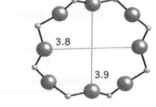

8-ring viewed along [110]   *8-ring viewed along [001]*

References:
(1)   Parnham, E.R. and Morris, R.E. *J. Am. Chem. Soc.*, **128**, 2204–2205 (2006)

SOD

Framework Type Data

Im3̄m

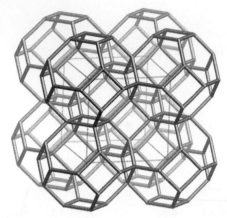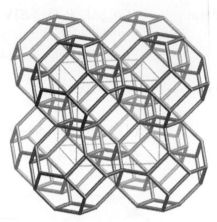

framework viewed along [100]

Idealized cell data: cubic, $Im\bar{3}m$, $a = 9.0$Å

Coordination sequences and vertex symbols:

| T_1 (12, $\bar{4}2m$) | 4 | 10 | 20 | 34 | 52 | 74 | 100 | 130 | 164 | 202 | | 4·4·6·6·6·6 |

Secondary building units: 6

Framework description: ABC sequence of 6-rings

Composite building units: *sod*

Materials with this framework type:

*Sodalite[1,2]
[Al-Co-P-O]-**SOD**[3]
[Al-Ga-Si-Ge-O]-**SOD**[4]
[Al-Ge-O]-**SOD**[5,6]
[Be-Al-Si-O]-**SOD**[7]
[Be-As-O]-**SOD**[8]
[Be-Ge-O]-**SOD**[9]
[Be-P-O]-**SOD**[8]
[Be-Si-O]-**SOD**[9,10]
[Co-Ga-P-O]-**SOD**[3,11]
[Ga-Ge-O]-**SOD**[6]

[Ga-Si-O]-**SOD**[12]
[Zn-Al-As-O]-**SOD**[13]
[Zn-As-O]-**SOD**[14]
[Zn-Ga-As-O]-**SOD**[11,13]
[Zn-Ga-P-O]-**SOD**[11]
[Zn-P-O]-**SOD**[14]
[Zn-Si-O]-**SOD**[15]
|Ca$_8$(WO$_4$)$_2$|[Al$_{12}$O$_{24}$]-**SOD**[16]
AlPO-20 plus variants[17,18]
Basic sodalite[19,20]
Bicchulite[21]

Danalite[22]
G[23]
Genthelvite[24]
Hauyn[25]
Helvin[22]
Hydroxo sodalite[26]
Nosean[27]
SIZ-9[28]
Silica sodalite[29]
TMA sodalite[30]
Tugtupite[31,32]

Sodalite — Type Material Data — **SOD**

Crystal chemical data: |Na$_8$Cl$_2$| [Al$_6$Si$_6$O$_{24}$]-**SOD**
cubic, $P\bar{4}3n$, a = 8.870Å [2]

Framework density: 17.2 T/1000Å^3

Channels: apertures formed by 6-rings only

References:
(1) Pauling, L. Z. *Kristallogr.*, **74**, 213–225 (1930)
(2) Loens, J. and Schulz, H. *Acta Crystallogr.*, **23**, 434–436 (1967)
(3) Feng, P., Bu, X. and Stucky, G.D. *Nature*, **388**, 735–741 (1997)
(4) Johnson, G.M., Mead, P.J. and Weller, M.T. *Microporous Mesoporous Mat.*, **38**, 445–460 (2000)
(5) Wiebcke, M., Sieger, P., Felsche, J., Engelhardt, G., Behrens, P. and Schefer, J. *Z. anorg. allg. Chemie*, **619**, 1321–1329 (1993)
(6) Bu, X., Feng, P., Gier, T.E., Zhao, D. and Stucky, G.D. *J. Am. Chem. Soc.*, **120**, 13389–13397 (1998)
(7) Armstrong, J.A. and Weller, M.T. *J. Chem. Soc., Dalton Trans.*, 2998–3005 (2006)
(8) Gier, T.E., Harrison, W.T.A. and Stucky, G.D. *Angew. Chem., Int. Ed.*, **30**, 1169–1171 (1991)
(9) Dann, S.E., Weller, M.T., Rainford, B.D. and Adroja, D.T. *Inorg. Chem.*, **36**, 5278–5283 (1997)
(10) Dann, S.E. and Weller, M.T. *Inorg. Chem.*, **35**, 555–558 (1996)
(11) Bu, X., Gier, T.E., Feng, P. and Stucky, G.D. *Microporous Mesoporous Mat.*, **20**, 371–379 (1998)
(12) McCusker, L.B., Meier, W.M., Suzuki, K. and Shin, S. *Zeolites*, **6**, 388–391 (1986)
(13) Feng, P., Zhang, T. and Bu, X. *J. Am. Chem. Soc.*, **123**, 8608–8609 (2001)
(14) Nenoff, T.M., Harrison, W.T.A., Gier, T.E. and Stucky, G.D. *J. Am. Chem. Soc.*, **113**, 378–379 (1991)
(15) Camblor, M.A., Lobo, R.F., Koller, H. and Davis, M.E. *Chem. Mater.*, **6**, 2193–2199 (1994)
(16) Depmeier, W. *Acta Crystallogr.*, **C40**, 226–231 (1984)
(17) Wilson, S.T., Lok, B.M., Messina, C.A., Cannan, T.R. and Flanigen, E.M. *J. Am. Chem. Soc.*, **104**, 1146–1147 (1982)
(18) Flanigen, E.M., Lok, B.M., Patton, R.L. and Wilson, S.T. *Proc. 7th Int. Zeolite Conf.*, pp. 103–112 (1986)
(19) Barrer, R.M. and White, E.A.D. *J. Chem. Soc.*, 1267–1278 (1951)
(20) Hassan, I. and Grundy, H.D. *Acta Crystallogr.*, **C39**, 3–5 (1983)
(21) Sahl, K. and Chatterjee, N.D. *Z. Kristallogr.*, **146**, 35–41 (1977)
(22) Glass, J.J., Jahns, R.H. and Stevens, R.E. *Am. Mineral.*, **29**, 163–191 (1944)
(23) Shishakova, T.N. and Dubinin, M.M. *Izv. Akad. Nauk SSSR*, 1303– (1965)
(24) Merlino, S. *Feldspars and Feldspathoids*, pp. 435–470 (1983)
(25) Loehn, J. and Schulz, H. *N. Jb. Miner. Abh.*, **109**, 201–210 (1968)
(26) Felsche, J., Luger, S. and Baerlocher, Ch. *Zeolites*, **6**, 367–372 (1986)
(27) Schulz, H. and Saalfeld, H. *Tschermaks Min. Petr. Mitt.*, **10**, 225–232 (1965)
(28) Parnham, E.R. and Morris, R.E. *J. Am. Chem. Soc.*, **128**, 2204–2205 (2006)
(29) Bibby, D.M. and Dale, M.P. *Nature*, **317**, 157–158 (1985)
(30) Baerlocher, Ch. and Meier, W.M. *Helv. Chim. Acta*, **52**, 1853–1860 (1969)
(31) Sorensen, H. *Am. Mineral.*, **48**, 1178 (1963)
(32) Hassan, I. and Grundy, H.D. *Can. Mineral.*, **29**, 385–390 (1991)

SOS

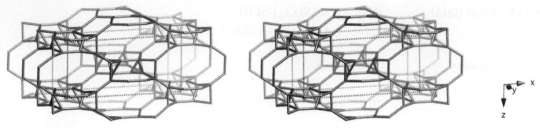

framework viewed along [010]

Idealized cell data: orthorhombic, *Pmna*, a = 20.4Å, b = 7.1Å, c = 9.7Å

Coordination sequences and vertex symbols:

T_1 (8,1)	4	10	19	31	47	71	97	124	153	187	231	278		$4·8·4·8·8_4·12$
T_2 (8,1)	4	8	15	27	46	70	93	116	151	197	235	265		$3·8_7·4·6_2·6_2·8$
T_3 (8,1)	4	7	15	27	41	68	93	119	146	192	230	268		$3·4·3·8_2·6_2·*$

Secondary building units: 4-2

Materials with this framework type:
  *SU-16[1]
  FJ-17[2]

SU-16 **Type Material Data** **SOS**

Crystal chemical data: |(C$_4$H$_{15}$N$_3$)$_4$| [B$_8$Ge$_{16}$O$_{48}$]-**SOS**
C$_4$H$_{13}$N$_3$ = diethylenetriamine
monoclinic, *P*2$_1$/*c*

$a = 6.936$Å, $b = 10.493$Å, $c = 20.448$Å, $\beta = 90.09°$ [1]

Stability: Unstable to removal of template [1]

Framework density: 16.1 T/1000Å^3

Channels: {[100] **12** 3.9 x 9.1 ↔ [010] **8** 3.3 x 3.3}**

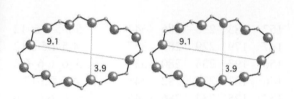 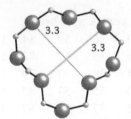 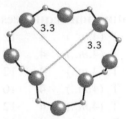

12-ring viewed along [100]   *8-ring viewed along [010]*

References:
(1) Li, Y. and Zou, X. *Angew. Chem., Int. Ed.*, **44**, 2012–2015 (2005)
(2) Zhang, H.-X., Zhang, J., Zheng, S.-T. and Yang, G.-Y. *Inorg. Chem.*, **44**, 1166–1168 (2005)

SSY

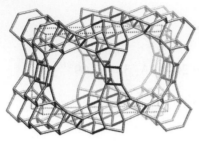 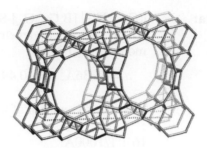

framework viewed along [100]

Idealized cell data: orthorhombic, *Pmmn* (origin choice 2), $a = 5.3$Å, $b = 22.6$Å, $c = 14.0$Å

Coordination sequences and vertex symbols:

T_1 (4,*m*..)	4	12	20	36	59	84	115	152	183	236	286	344	$5 \cdot 5 \cdot 5 \cdot 5 \cdot 6_2 \cdot 12_2$
T_2 (4,*m*..)	4	10	19	35	57	84	113	147	179	229	289	337	$4 \cdot 5 \cdot 4 \cdot 5 \cdot 6 \cdot 12_2$
T_3 (4,*m*..)	4	12	24	38	57	79	116	156	193	233	280	330	$5_2 \cdot 6_2 \cdot 6 \cdot 6_2 \cdot 6 \cdot 6_2$
T_4 (4,*m*..)	4	10	20	35	57	83	115	146	183	227	282	348	$4 \cdot 5 \cdot 4 \cdot 5 \cdot 12_2 \cdot *$
T_5 (4,*m*..)	4	12	22	37	57	87	117	152	185	234	286	345	$5_2 \cdot 6_2 \cdot 6 \cdot 6_2 \cdot 6 \cdot 6_2$
T_6 (4,*m*..)	4	12	22	37	55	87	120	148	180	234	294	344	$5 \cdot 5 \cdot 5 \cdot 5 \cdot 6_2 \cdot *$
T_7 (2,*mm*2)	4	12	24	44	56	78	110	158	206	234	262	336	$5 \cdot 5 \cdot 5 \cdot 5 \cdot 12_2 \cdot *$
T_8 (2,*mm*2)	4	12	22	38	60	78	114	150	200	232	280	332	$5 \cdot 5 \cdot 5 \cdot 5 \cdot 6_2 \cdot 12_2$

Secondary building units: see *Compendium*

Composite building units:

dzc	*jbw*	*mtt*	*ton*
double zigzag chain			

Materials with this framework type:
*SSZ-60[1]

SSZ-60   **Type Material Data**   **SSY**

Crystal chemical data: $[B_{0.6}Si_{27.4}O_{56}]$-**SSY**
orthorhombic, $Pmn2_1$, $a = 21.951$Å, $b = 13.698$Å, $c = 5.012$Å [1]
    (Space group changed to standard setting)
(Relationship to unit cell of Framework Type: $a' = b$, $b' = c$, $c' = a$)

Framework density: 18.6 T/1000Å^3

Channels: [001] **12** 5.0 x 7.6*

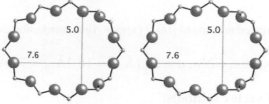

12-ring viewed along [001]

References:

(1)   Burton, A. and Elomari, S. *Chem. Commun.*, 2618–2619 (2004)

STF

Framework Type Data

C2/m

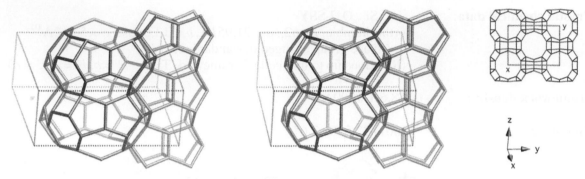

framework viewed normal to [001] (upper right: projection down [001])

Idealized cell data: monoclinic, $C2/m$, $a = 14.1$Å, $b = 18.2$Å, $c = 7.5$Å, $\beta = 99°$

Coordination sequences and vertex symbols:

T_1 (8,1)	4	12	20	34	56	88	114	143	173	224	$5 \cdot 5_2 \cdot 5 \cdot 6 \cdot 5 \cdot 6$
T_2 (8,1)	4	11	22	39	55	82	111	149	188	223	$4 \cdot 5 \cdot 5 \cdot 6 \cdot 5 \cdot 10$
T_3 (8,1)	4	11	23	36	59	80	113	147	183	227	$4 \cdot 5 \cdot 5 \cdot 6 \cdot 5 \cdot 10$
T_4 (4,m)	4	11	19	36	54	84	110	146	179	226	$4 \cdot 6 \cdot 5 \cdot 5 \cdot 5 \cdot 5$
T_5 (4,m)	4	11	20	31	58	84	117	137	174	229	$4 \cdot 6_2 \cdot 5 \cdot 5 \cdot 5 \cdot 5$

Secondary building units: 5-3

Composite building units:

cas

stf

Materials with this framework type:
*SSZ-35[1]
ITQ-9[2]
Mu-26[3]

SSZ-35, as synthesized[4]
STF-SFF structural intermediates[5]

SSZ-35   **Type Material Data**   **STF**

Crystal chemical data: [$Si_{16}O_{32}$]-**STF**

triclinic, $P\bar{1}$, $a = 11.411$Å, $b = 11.527$Å, $c = 7.377$Å

$\alpha = 94.66\bullet$, $\beta = 96.21$, $\gamma = 104.89°$ [1]

(Relationship to unit cell of Framework Type: $V' = V/2$)

Framework density: 17.3 T/1000Å³

Channels: [001] **10** 5.4 x 5.7*

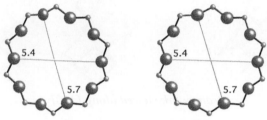

10-ring viewed along [001]

References:

(1) Wagner, P., Zones, S.I., Davis, M.E. and Medrud, R.C. *Angew. Chem., Int. Ed.*, **38**, 1269–1272 (1999)

(2) Villaescusa, L.A., Barrett, P.A. and Camblor, M.A. *Chem. Commun.*, **21**, 2329–2330 (1998)

(3) Harbuzaru, B., Roux, M., Paillaud, J.L., Porcher, F., Marichal, C., Chezeau, J.M. and Patarin, J. *Chemistry Letters*, 616–617 (2002)

(4) Fyfe, C.A., Brouwer, D.H., Lewis, A.R., Villaescusa, L.A. and Morris, R.E. *J. Am. Chem. Soc.*, **124**, 7770–7778 (2002)

(5) Villaescusa, L.A., Zhou, W., Morris, R.E. and Barrett, P.A. *J. Mater. Chem.*, **14**, 1982–1987 (2004)

STI

Framework Type Data

Fmmm

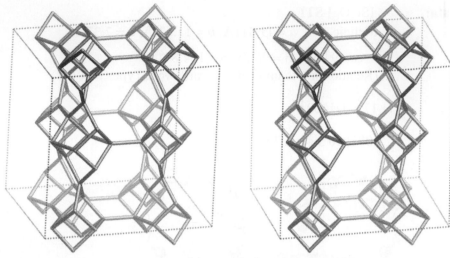

framework viewed along [100]

Idealized cell data: orthorhombic, *Fmmm*, $a = 13.5\text{Å}$, $b = 17.8\text{Å}$, $c = 17.9\text{Å}$

Coordination sequences and vertex symbols:

T_1 (32,1)	4	10	20	34	57	82	103	138	181	220	$4 \cdot 5 \cdot 4 \cdot 6 \cdot 5 \cdot 6_2$
T_2 (16,m)	4	11	20	36	57	78	109	140	176	222	$4 \cdot 8 \cdot 5 \cdot 8_2 \cdot 5 \cdot 8_2$
T_3 (16,m)	4	9	17	35	57	77	103	138	188	225	$4 \cdot 5 \cdot 4 \cdot 5 \cdot 4 \cdot 8$
T_4 (8,222)	4	12	18	34	58	82	112	130	172	228	$5_2 \cdot 5_2 \cdot 6 \cdot 6 \cdot 10 \cdot 10$

Secondary building units: 4-4=1

Composite building units:

sti   *bre*

Materials with this framework type:

*Stilbite[1-3]

Barrerite[4]

Stellerite[5]

Synthetic barrerite[6]

Synthetic stellerite[6]

Synthetic stilbite[7]

TNU-10[8]

Stilbite   **Type Material Data**   **STI**

Crystal chemical data: $|Na_4Ca_8 (H_2O)_{56}| [Al_{20}Si_{52}O_{144}]$-**STI**
monoclinic, $C2/m$

$a = 13.64Å, b = 18.24Å, c = 11.27Å, \beta = 128.0°$ [3]
(Relationship to unit cell of Framework Type:

$a' = a, b' = b, c' = c/(2\sin\beta')$
or, as vectors, **a' = a, b' = b, c' = (c - a)/2**)

Framework density: 16.3 T/1000Å^3

Channels: [100] **10** 4.7 x 5.0* ↔ [001] **8** 2.7 x 5.6*

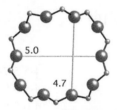 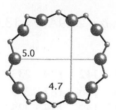 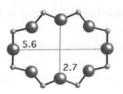 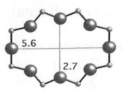

10-ring viewed along [100]   *8-ring along [001]*

References:
(1) Galli, E. and Gottardi, G. *Miner. Petrogr. Acta*, **12**, 1–10 (1966)
(2) Slaughter, M. *Am. Mineral.*, **55**, 387–397 (1970)
(3) Galli, E. *Acta Crystallogr.*, **B27**, 833–841 (1971)
(4) Galli, E. and Alberti, A. *Bull. Soc. fr. Mineral. Cristallogr.*, **98**, 331–340 (1975)
(5) Galli, E. and Alberti, A. *Bull. Soc. fr. Minéral. Cristallogr.*, **98**, 11–18 (1975)
(6) Ghobarkar, H., Schaef, O. and Guth, U. *J. Solid State Chem.*, **142**, 451–454 (1999)
(7) Ghobarkar, H. and Schaef, O. *J. Phys. D: Appl. Phys.*, **31**, 3172–3176 (1998)
(8) Hong, S.B., Lear, E.G., Wright, P.A., Zhou, W., Cox, P.A., Shin, C.-H., Park, J.-H. and Nam. I.-S. *J. Am. Chem. Soc.*, **126**, 5817–5826 (2004)

STT

Framework Type Data

$P2_1/n$

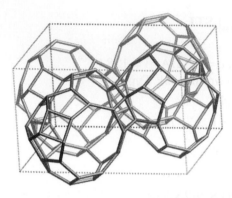
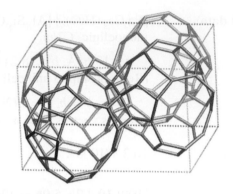

framework viewed normal to [100]

Idealized cell data: monoclinic, $P2_1/n$, $a = 13.1$Å, $b = 21.9$Å, $c = 13.6$Å, $\beta = 102.9°$

Coordination sequences and vertex symbols:

T_1 (4,1)	4	11	22	36	55	84	111	142	179	233	$4·5·5·7·6·9$
T_2 (4,1)	4	10	23	37	57	78	108	146	187	225	$4·4·5·9·6·7$
T_3 (4,1)	4	11	20	35	54	84	108	142	178	225	$4·6·5·6·5·7$
T_4 (4,1)	4	10	19	32	55	81	109	141	174	223	$4·5·4·6·5·5$
T_5 (4,1)	4	9	21	38	57	78	104	144	195	220	$4·4·4·9·5·7$
T_6 (4,1)	4	10	20	40	57	78	105	143	191	226	$4·5·4·7·5·9$
T_7 (4,1)	4	10	20	35	57	77	108	139	184	230	$4·6·4·6·5·7$
T_8 (4,1)	4	10	19	35	54	77	108	146	174	215	$4·5·4·9·5·6$
T_9 (4,1)	4	11	19	34	53	81	115	140	173	219	$4·6·5·5·5·5$
T_{10} (4,1)	4	10	21	31	51	83	115	134	172	216	$4·6·4·6·5·5$
T_{11} (4,1)	4	12	20	31	50	88	117	137	167	221	$5·5_2·5·6·6·6$
T_{12} (4,1)	4	10	18	35	59	77	107	150	179	221	$4·5·4·9·5·5$
T_{13} (4,1)	4	10	20	38	55	75	104	149	185	218	$4·4·5·6·5·9$
T_{14} (4,1)	4	12	21	33	58	86	112	141	174	226	$5·5·5·6·5·7$
T_{15} (4,1)	4	11	21	33	55	78	111	145	175	217	$4·5·5·6·5·9$
T_{16} (4,1)	4	11	20	32	54	89	115	138	172	221	$4·6_2·5·5·5·5$

Secondary building units: 5-3

Composite building units:

bea   *cas*

Materials with this framework type:
   *SSZ-23[1]

SSZ-23   **Type Material Data**   **STT**

Crystal chemical data: $|(C_{13}H_{24}N)_{4.1} F_{3.3}(OH)_{0.8}| [Si_{64}O_{128}]$-**STT**
$C_{13}H_{24}N$ = N,N,N-trimethyl-1-adamantammonium
monoclinic, $P2_1/n$

$a = 12.959$Å, $b = 21.792$Å, $c = 13.598$Å, $\beta = 101.85°$ [1]

Framework density: 17 T/1000Å^3

Channels: [101] **9** 3.7 x 5.3* $\leftrightarrow$ [001] **7** 2.4 x 3.5*

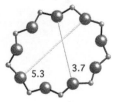 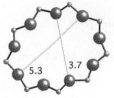   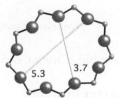 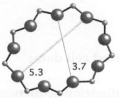

      9-ring viewed along [101]   *7-ring viewed along [001]*

References:
(1) Camblor, M.A., Diaz-Cabanas, M.-J., Perez-Pariente, J., Teat, S.J., Clegg, W., Shannon, I.J., Lightfoot, P., Wright, P.A. and Morris, R.E. *Angew. Chem., Int. Ed.*, **37**, 2122–2126 (1998)

SZR

Framework Type Data

Cmmm

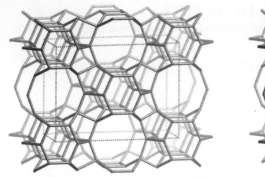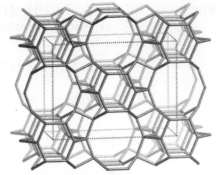

framework viewed along [001]

Idealized cell data: orthorhombic, *Cmmm*, $a = 18.9$Å, $b = 14.4$Å, $c = 7.5$Å

Coordination sequences and vertex symbols:

T_1 (16,1)	4	10	21	38	61	88	116	149	190	237	289	340	$4 \cdot 5 \cdot 4 \cdot 6 \cdot 6 \cdot 8_2$
T_2 (8,..*m*)	4	12	23	38	59	82	119	155	191	231	283	338	$5 \cdot 8 \cdot 5 \cdot 8 \cdot 5_2 \cdot 6$
T_3 (8,.*m*.)	4	10	19	34	60	89	118	144	179	238	296	342	$4 \cdot 5 \cdot 4 \cdot 5 \cdot 6 \cdot 8_2$
T_4 (4,2*mm*)	4	12	24	38	52	88	118	150	192	232	278	340	$5 \cdot 5 \cdot 5 \cdot 5 \cdot 10_2 \cdot 10_4$

Secondary building units: 6

Composite building units:

mtt   *d6r*   *mso*

Materials with this framework type:
 *SUZ-4[1,2]

SUZ-4   Type Material Data   **SZR**

Crystal chemical data: $|K_4|$ $[Al_4Si_{32}O_{72}]$-**SZR**
orthorhombic, *Cmmm*, $a = 18.8064$Å, $b = 14.2298$Å, $c = 7.4548$Å [2]

Framework density: 18 T/1000Å^3

Channels: { [001] **10** 4.1 x 5.2 ↔ [010] **8** 3.2 x 4.8 ↔ [110] **8** 3.0 x 4.8 }***

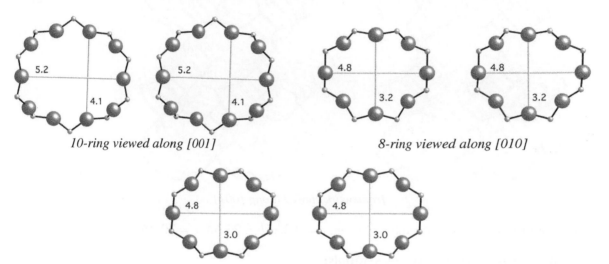

10-ring viewed along [001]   *8-ring viewed along [010]*

8-ring viewed along [110]

References:
(1) Lawton, S.L., Bennett, J.M., Schlenker, J.L. and Rubin, M.K. *Chem. Commun.*, 894–896 (1993)
(2) Strohmaier, K.G., Afeworki, M. and Dorset, D.L. *Z. Kristallogr.*, **221**, 689–698 (2006)

TER

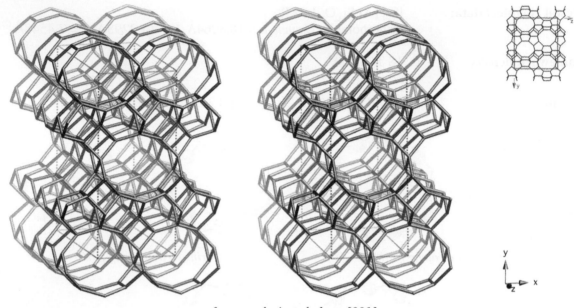

framework viewed along [001]

Idealized cell data: orthorhombic, *Cmcm*, $a = 9.8$Å, $b = 23.6$Å, $c = 20.2$Å

Coordination sequences and vertex symbols:

T_1 (16,1)	4	11	19	35	57	83	113	137	184	231	$4 \cdot 5 \cdot 5 \cdot 6 \cdot 5 \cdot 6_2$
T_2 (16,1)	4	11	21	35	58	87	103	144	188	227	$4 \cdot 5 \cdot 5 \cdot 6_2 \cdot 5 \cdot 10_3$
T_3 (8,m)	4	11	22	39	62	82	104	142	178	225	$4 \cdot 10_6 \cdot 5 \cdot 6_3 \cdot 5 \cdot 6_3$
T_4 (8,m)	4	11	21	41	61	77	107	134	186	232	$4 \cdot 5_2 \cdot 5 \cdot 10_4 \cdot 5 \cdot 10_4$
T_5 (8,m)	4	11	19	34	56	78	116	152	184	208	$4 \cdot 5_2 \cdot 5 \cdot 6 \cdot 5 \cdot 6$
T_6 (8,m)	4	11	21	33	53	80	112	155	187	214	$4 \cdot 10_2 \cdot 5 \cdot 6_3 \cdot 5 \cdot 6_3$
T_7 (8,m)	4	12	20	35	55	81	119	142	182	216	$5 \cdot 6 \cdot 5 \cdot 6 \cdot 5_2 \cdot 10_2$
T_8 (8,m)	4	12	23	32	49	86	124	147	167	219	$5 \cdot 6_2 \cdot 5 \cdot 6_2 \cdot 10 \cdot 10_4$

Secondary building units: 2-6-2 or 4-1

Composite building units:

    bre   *bog*   *cas*

Materials with this framework type:

    *Terranovaite[1]

Terranovaite   Type Material Data   **TER**

Crystal chemical data: $|Na_{4.2}K_{0.2}Mg_{0.2}Ca_{3.7}(H_2O)_{29}|\ [Al_{12.3}Si_{67.7}O_{160}]$-**TER**
orthorhombic, *Cmcm*, $a = 9.747$Å, $b = 23.880$Å, $c = 20.068$Å [1]

Framework density: 17.1 T/1000Å^3

Channels: [100] **10** 5.0 x 5.0* $\leftrightarrow$ [001] **10** 4.1 x 7.0*

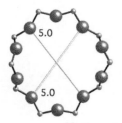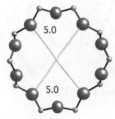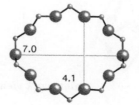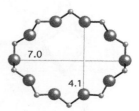

10-ring viewed along [100]   *10-ring viewed along [001]*

References:
(1)   Galli, E., Quartieri, S., Vezzalini, G., Alberti, A. and Franzini, M. *Am. Mineral.*, **82**, 423–429 (1997)

THO

Framework Type Data

Pmma

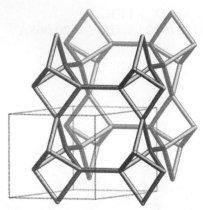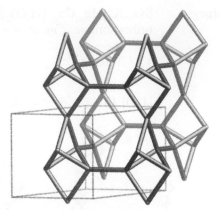

framework viewed normal to [001]

Idealized cell data: orthorhombic, *Pmma*, $a = 14.0$Å, $b = 7.0$Å, $c = 6.5$Å

Coordination sequences and vertex symbols:

T_1 (4,*m*)	4	9	19	35	52	72	100	131	163	201		$4 \cdot 8_3 \cdot 4 \cdot 8_3 \cdot 4_2 \cdot 8_4$
T_2 (4,*m*)	4	9	19	33	50	74	100	129	165	201		$4 \cdot 8_3 \cdot 4 \cdot 8_3 \cdot 4_2 \cdot 8_4$
T_3 (2,*mm*2)	4	8	18	34	50	68	100	130	160	204		$4_2 \cdot 4_2 \cdot 8_4 \cdot 8_4 \cdot 8_4 \cdot 8_4$

Secondary building units: 4=1

Composite building units:

nat

Materials with this framework type:

*Thomsonite[1-3]

[Al-Co-P-O]-**THO**[4]

[Ga-Co-P-O]-**THO**[4]

[Zn-Al-As-O]-**THO**[5]

[Zn-P-O]-**THO**[6]

[Zn-P-O]-**THO**[7]

|Rb$_{20}$|[Ga$_{20}$Ge$_{20}$O$_{80}$]-**THO**[8]

Na-V ([Ga-Si-O]-**THO**)[9]

Synthetic thomsonite[10]

ZCP-**THO** ([Zn-Co-P-O]-**THO**)[11]

Thomsonite Type Material Data **THO**

Crystal chemical data: $|Na_4Ca_8(H_2O)_{24}|$ $[Al_{20}Si_{20}O_{80}]$-**THO**
orthorhombic, *Pncn*, $a = 13.088$Å, $b = 13.052$Å, $c = 13.229$Å [3]
(Relationship to unit cell of Framework Type: $a' = a$, $b' = 2b$, $c' = 2c$)

Framework density: 17.7 T/1000Å^3

Channels: [100] **8** 2.3 x 3.9* $\leftrightarrow$ [010] **8** 2.2 x 4.0* $\leftrightarrow$ [001] **8** 2.2 x 3.0*
(variable due to considerable flexibility of framework)

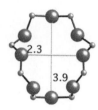 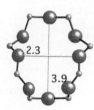 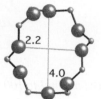 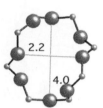

8-ring viewed along [100] *8-ring viewed along [010]*

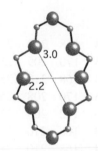 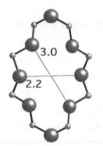

8-ring along [001], variable

References:
(1) Taylor, W.H., Meek, C.A. and Jackson, W.W. *Z. Kristallogr.*, **84**, 373–398 (1933)
(2) Alberti, A., Vezzalini, G. and Tazzoli, V. *Zeolites*, **1**, 91–97 (1981)
(3) Pluth, J.J., Smith, J.V. and Kvick, A. *Zeolites*, **5**, 74–80 (1985)
(4) Feng, P., Bu, X. and Stucky, G.D. *Nature*, **388**, 735–741 (1997)
(5) Feng, P., Zhang, T. and Bu, X. *J. Am. Chem. Soc.*, **123**, 8608–8609 (2001)
(6) Neeraj, S. and Natarajan, S. *J. Phys. Chem. Solids*, **62**, 1499–1505 (2001)
(7) Ng, H.Y. and Harrison, W.T.A. *Microporous Mesoporous Mat.*, **50**, 187–194 (2001)
(8) Lee, Y.J., Kim, S.J. and Parise, J.B. *Microporous Mesoporous Mat.*, **34**, 255–271 (2000)
(9) Barrer, R.M., Baynham, J.W., Bultitude, F.W. and Meier, W.M. *J. Chem. Soc.*, 195–208 (1959)
(10) Ghobarkar, H. and Schaef, O. *Cryst. Res. Technol.*, **32**, 653–657 (1997)
(11) Ke, Y.X., He, G.F., Li, J.M., Zhang, Y.G. and Lu, S.M. *New J. Chem.*, **25**, 1627–1630 (2001)

TOL

Framework Type Data

$P\bar{3}m1$

framework viewed normal to [001] (upper right: projection down [001])

Idealized cell data: trigonal, $P\bar{3}m1$, $a = 12.3$Å, $c = 30.9$Å

Coordination sequences and vertex symbols:

T_1 (12,1)	4	10	20	34	54	78	104	134	168	210	256	302		4·4·6·6·6·6
T_2 (12,1)	4	10	20	34	53	76	103	135	170	209	252	300		4·6·4·6·6·6
T_3 (12,1)	4	10	20	34	54	78	104	134	168	210	256	302		4·6·4·6·6·6
T_4 (12,1)	4	10	20	34	53	76	103	135	170	208	250	299		4·6·4·6·6·6
T_5 (12,1)	4	10	20	34	52	74	102	136	172	209	250	299		4·6·4·6·6·6
T_6 (6,.2.)	4	10	20	34	54	78	104	134	168	210	256	302		4·4·6·6·6·6
T_7 (6,.2.)	4	10	20	34	54	78	104	134	168	210	256	302		4·4·6·6·6·6

Secondary building units: 4 or 6

Framework description: CACACBCBCACB sequence of 6-rings

Composite building units:

can   *los*   *lio*

Materials with this framework type:

*Tounkite-like mineral[1]

Tounkite-like mineral   Type Material Data   **TOL**

Crystal chemical data: $|Na_{31.1}Ca_{15.94}K_{0.96}Cl_8(SO_4)_{9.3}(SO_3)_{0.7}|$ $[Al_{36}Si_{36}O_{144}]$-**TOL**
trigonal, $P3$, $a = 12.757$Å, $c = 32.211$Å $^{(1)}$

Framework density: 15.9 T/1000Å^3

Channels: apertures formed by 6-rings only

References:
(1) Rozenberg, K.A., Sapozhnikov, A.N., Rastsvetaeva, R.K., Bolotina, N.B. and Kashaev, A.A. *Crystallogr. Reports*, **4**, 635–642 (2004)

TON

Framework Type Data

Cmcm

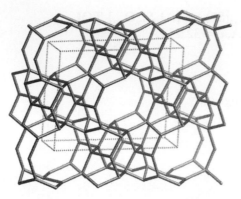
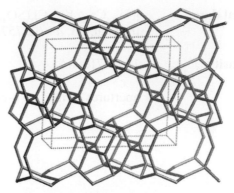

framework viewed along [001]

Idealized cell data: orthorhombic, *Cmcm*, $a = 14.1$Å, $b = 17.8$Å, $c = 5.3$Å

Coordination sequences and vertex symbols:

T_1 (8,*m*)	4	12	24	40	61	96	133	163	204	262	$5_2 \cdot 6_2 \cdot 6 \cdot 6_2 \cdot 6 \cdot 6_2$
T_2 (8,*m*)	4	12	23	43	66	91	128	169	214	258	$5 \cdot 5 \cdot 5 \cdot 5 \cdot 6 \cdot 10_2$
T_3 (4,*m2m*)	4	12	22	41	68	97	118	166	224	258	$5 \cdot 5 \cdot 5 \cdot 5 \cdot 6_2 \cdot 10_2$
T_4 (4,*m2m*)	4	12	22	39	66	95	130	158	208	270	$5 \cdot 5 \cdot 5 \cdot 5 \cdot 6_2 \cdot *$

Secondary building units: 5-1

Composite building units:

jbw	mtt	bik	ton

Materials with this framework type:

*Theta-1[1-3]
ISI-1[4]
KZ-2[5]

NU-10[6]
ZSM-22[7,8]

Theta-1

Type Material Data

TON

Crystal chemical data: $|Na_n (H_2O)_4|$ $[Al_nSi_{24-n}O_{48}]$-**TON**, n < 2
orthorhombic, $Cmc2_1$, $a = 13.859$Å, $b = 17.420$Å, $c = 5.038$Å [8]

Framework density: 19.7 T/1000Å^3

Channels: [001] **10** 4.6 x 5.7*

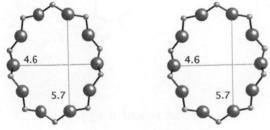

10-ring viewed along [001]

References:
(1) Barri, S.A.I., Smith, G.W., White, D. and Young, D. *Nature*, **312**, 533–534 (1984)
(2) Highcock, R.M., Smith, G.W. and Wood, D. *Acta Crystallogr.*, **C41**, 1391–1394 (1985)
(3) Papiz, Z., Andrews, S.J., Damas, A.M., Harding, M.M. and Highcock, R.M. *Acta Crystallogr.*, **C46**, 172–173 (1990)
(4) Kozo, T. and Noboru, K. *E. Patent A-170,003* (1986)
(5) Parker, L.M. and Bibby, D.M. *Zeolites*, **3**, 8–11 (1983)
(6) Araya, A. and Lowe, B.M. *Zeolites*, **4**, 280–286 (1984)
(7) Kokotailo, G.T., Schlenker, J.L., Dwyer, F.G. and Valyocsik, E.W. *Zeolites*, **5**, 349–351 (1985)
(8) Marler, B. *Zeolites*, **7**, 393–397 (1987)

TSC

Framework Type Data

$Fm\bar{3}m$

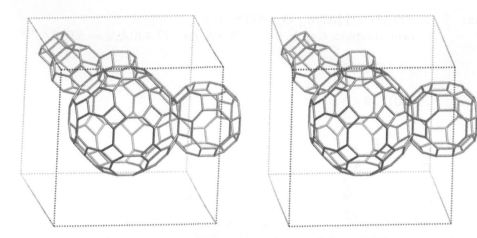

framework viewed along [001]

Idealized cell data: cubic, $Fm\bar{3}m$, $a = 30.7$Å

Coordination sequences and vertex symbols:

T_1 (192,1)	4	9	16	25	37	53	74	99	125	151	4·4·4·6·6·8
T_2 (192,1)	4	9	17	28	41	56	73	93	117	146	4·4·4·8·6·8

Secondary building units: 8-8 or 8 or 6-6 or 6 or 4-2 or 4

Composite building units:

d6r	*d8r*	*sod*	*lta*

Materials with this framework type:
  *Tschörtnerite[1]

Tschörtnerite   Type Material Data   TSC

Crystal chemical data: $|Ca_{64}(K_2,Ca,Sr,Ba)_{48}Cu_{48}(H_2O)_x(OH)_{128}|\,[Al_{192}Si_{192}O_{768}]$-**TSC**
cubic, $Fm\overline{3}m$, $a = 31.62\text{Å}$ [1]

Framework density: $12.1\ T/1000\text{Å}^3$

Channels: <100> **8** 4.2 x 4.2*** ↔ <110> **8** 3.1 x 5.6***

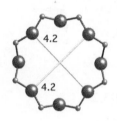 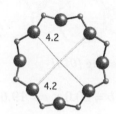   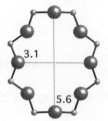 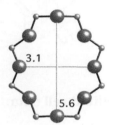

8-ring viewed along <100>   *8-ring viewed along <110>*

References:
(1) Effenberger, H., Giester, G., Krause, W. and Bernhardt, H.J. *Am. Mineral.*, **83**, 607–617 (1998)

TUN

Framework Type Data

C2/m

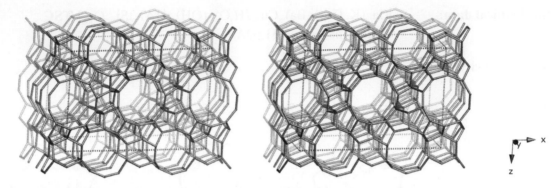

framework viewed along [010]

Idealized cell data: monoclinic, $C2/m$, $a = 27.8$Å, $b = 20.0$Å, $c = 19.6$Å, $\beta = 93.2°$

Coordination sequences and vertex symbols:
   see Appendix A for a list of the coordination sequences and vertex symbols for the 24 T-atoms

Secondary building units: 5-1

Composite building units:

| *mor* | *mtt* | *cas* | *lau* | *stf* |

Materials with this framework type:
   *TNU-9[1]

TNU-9   **Type Material Data**   **TUN**

Crystal chemical data: |H$_{9.3}$| [Al$_{9.3}$Si$_{182.7}$O$_{384}$]-**TUN**
                              monoclinic, *C2/m*

                              $a = 28.2219$ Å, $b = 20.0123$ Å, $c = 19.4926$ Å, β = 92.33 °$^{(1)}$

Framework density: 17.5 T/1000Å^3

Channels: {[010] **10** 5.6 x 5.5 ↔ [10$\overline{1}$] **10** 5.4 x 5.5}***
                              (There are 2 different channels along [010]; the smaller is 5.1 x 5.5 Å)

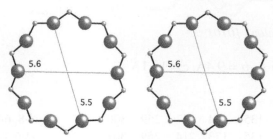

10-ring viewed along [010]

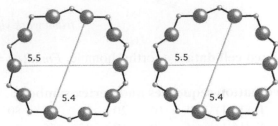

10-ring viewed along [10$\overline{1}$]

References:
(1) Gramm, F., Baerlocher, Ch., McCusker, L.B., Warrender, S.J., Wright, P.A., Han, B., Hong, S.B., Liu, Z., Ohsuna, T. and Terasaki, O. *Nature*, **444**, 79–81 (2006)

UEI

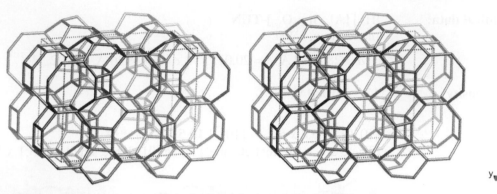

framework viewed along [010]

Idealized cell data: orthorhombic, *Fmm2*, $a = 19.5$ Å, $b = 9.4$Å, $c = 15.1$Å

Coordination sequences and vertex symbols:

T_1 (16,1)	4	10	20	35	56	80	105	135	174	217	259	306	$4 \cdot 6_2 \cdot 4 \cdot 8_3 \cdot 6 \cdot 8_2$
T_2 (16,1)	4	9	19	35	55	77	102	135	173	214	259	307	$4 \cdot 4 \cdot 4 \cdot 6 \cdot 8 \cdot 8_3$
T_3 (16,1)	4	10	21	36	53	76	108	142	173	210	259	310	$4 \cdot 6 \cdot 4 \cdot 8_2 \cdot 6 \cdot 8$

Secondary building units: 6 or 4-2 or 4

Composite building units:

dcc
double
crankshaft chain

Materials with this framework type:

*Mu-18[1]

Mu-18   **Type Material Data**   **UEI**

Crystal chemical data: $|(C_5H_{14}N_2)_4 (H_2O)_4| [Ga_{24}P_{24}O_{96} (OH)_8]$-**UEI**
$C_5H_{12}N_2$ = 1-methylpiperazine
orthorhombic, *Aea*2, a = 18.035Å, b = 10.513Å, c = 14.293Å [1]

Stability: reversibly adsorbs water below 350°C [1]

Framework density: 17.7 T/1000Å^3

Channels: {[010] **8** 3.5 x 4.6 ↔ [001] **8** 2.5 x 3.6}**

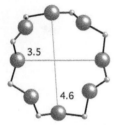 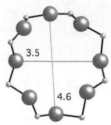   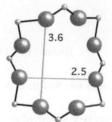 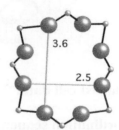

8-ring viewed along [010]   *8-ring viewed along [001]*

References:
(1) Josien, L., Simon, A., Gramlich, V. and Patarin, J. *Chem. Mater.*, **13**, 1305–1311 (2001)

UFI Framework Type Data *I4/mmm*

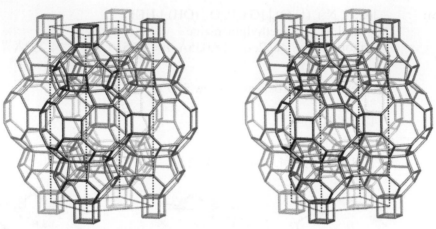

framework viewed along [100]

Idealized cell data: tetragonal, *I4/mmm*, $a = 12.1$Å, $c = 28.6$Å

Coordination sequences and vertex symbols:

T_1 (16,.m.)	4	9	18	33	51	71	97	124	153	195	241	283	4·6·4·6·4·8
T_2 (32,1)	4	10	19	31	46	64	91	123	156	194	231	271	4·4·5·6·5·8
T_3 (16,.m.)	4	10	19	30	50	70	93	124	155	196	230	280	4·6·4·6·5·8

Secondary building units: 8

Composite building units:

          d4r   *rth*   *lta*

  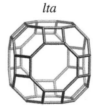

Materials with this framework type:
  *UZM-5[1]

UZM-5   **Type Material Data**   **UFI**

Crystal chemical data: $|K_8|$ $[Al_8Si_{56}O_{128}]$-**UFI**
tetragonal, $I4/mmm$, $a = 12.1507$Å, $c = 28.172$Å $^{(1)}$

Framework density: 15.4 T/1000Å^3

Channels: <100> **8** 3.6 x 4.4** $\leftrightarrow$ [001] **8** 3.2 x 3.2 (cage) (i.e. ends in a cage with this window to the cage)

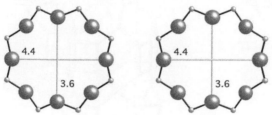

8-ring viewed along [100]

References:
(1)   Blackwell, C.S., Broach, R.W., Gatter, M.G., Holmgren, J.S., Jan, D.-Y., Lewis, G.J., Mezza, B.J., Mezza, T.M., Miller, M.A., Moscoso, J.G., Patton, R.L., Rohde, L.M., Schoonover, M.W., Sinkler, W., Wilson, B.A. and Wilson, S.T. *Angew. Chem., Int. Ed.*, **42**, 1737–1740 (2003)

UOZ

Framework Type Data

P4/nnc

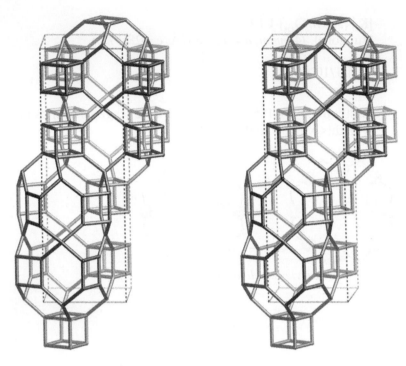

framework viewed along [010]

Idealized cell data: tetragonal, *P4/nnc* (origin choice 2), *a* = 8.6Å, *c* = 27.5Å

Coordination sequences and vertex symbols:

T_1 (16,1)	4	9	19	34	49	69	100	133	162	195	238	288	4·6·4·6·4·6
T_2 (16,1)	4	9	19	35	51	69	97	131	164	197	236	287	4·6·4·6·4·6
T_3 (4, $\overline{4}$)	4	12	18	28	52	82	100	120	162	208	244	274	6·6·6·6·6·6
T_4 (4,222)	4	12	18	26	52	80	88	110	162	214	244	268	$6·6·6_2·6_2·12_8·12_8$

Secondary building units: 4-1

Composite building units:

  d4r   *lau*

Materials with this framework type:
  *IM-10[1]

Crystal chemical data: $|(C_{12}N_2H_{30})_2\,F_4|\,[Ge_{40}O_{80}]$-**UOZ**
$C_{12}N_2H_{30}$ = hexamethonium = (1,6-bis(trimethylamino)hexane)
tetragonal, $P\bar{4}n2$, a = 9.1596Å, c = 28.5614Å [1]

Framework density: 16.7 T/1000Å^3

Channels: apertures formed by 6-rings only

References:
(1) Mathieu, Y., Paillaud, J.-L., Caullet, P. and Bats, N. *Microporous Mesoporous Mat.*, **75**, 13–22 (2004)

USI

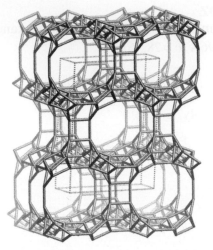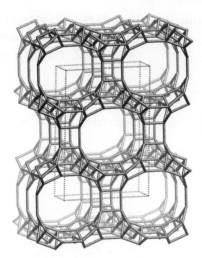

framework viewed along [001]

Idealized cell data: monoclinic, *C2/m*, $a = 21.1$Å, $b = 13.0$Å, $c = 9.7$Å, $\beta = 108.4°$

Coordination sequences and vertex symbols:

T_1 (8,1)	4	10	17	27	46	70	94	112	134	181	230	259	$4 \cdot 6_2 \cdot 4 \cdot 6_2 \cdot 10 \cdot 12_2$
T_2 (8,1)	4	9	15	26	43	61	86	114	139	174	217	260	$4 \cdot 6_3 \cdot 4 \cdot 6_3 \cdot 4 \cdot 12_2$
T_3 (8,1)	4	9	17	26	41	65	85	108	145	181	213	252	$4 \cdot 4 \cdot 4 \cdot 6 \cdot 6 \cdot 6_2$
T_4 (8,1)	4	10	19	32	47	60	84	122	153	177	209	255	$4 \cdot 4 \cdot 6_3 \cdot 10 \cdot 6_3 \cdot 10$
T_5 (8,1)	4	9	18	29	44	68	92	116	144	173	214	267	$4 \cdot 4 \cdot 4 \cdot 6 \cdot 6 \cdot 6$

Secondary building units: 4-1

Composite building units:

dzc	*mei*	*bog*
double zigzag chain		

Materials with this framework type:
  *IM-6[1]

Crystal chemical data: $|(C_5H_{13}N_2)_2 (H_3O)_2| [Co_4Ga_6P_{10}O_{40}]$-**USI**
$C_5H_{13}N_2 = $ 1-methylpiperazinium
triclinic, $P\bar{1}$, $a = $ 9.848Å, $b = $ 12.470Å, $c = $ 12.603Å

$\alpha = $ 63.47°, $\beta = $ 74.56°, $\gamma = $ 76.03° [1]

Framework density: 15.1 T/1000Å^3

Channels: [100] **12** 6.1 x 6.2 * ↔ [001] **10** 3.9 x 6.4*

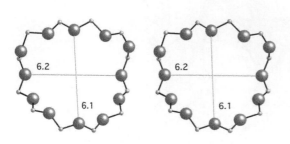
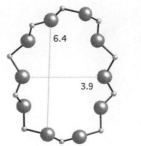
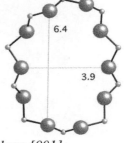

12-ring viewed along [100]   *10-ring viewed along [001]*

References:
(1)   Josien, L., Simon Masseron, A., Gramlich, V., Patarin, J. and Rouleau, L. *Chem. Eur. Journal*, **9**, 856–861 (2003)

UTL

Framework Type Data

C2/m

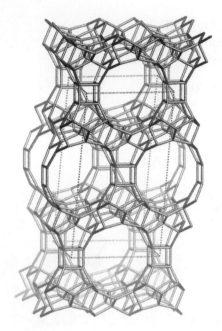
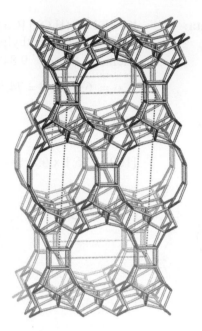

framework viewed along [001]

Idealized cell data: monoclinic, *C2/m*, *a* = 29.0Å, *b* = 14.0Å, *c* = 12.4Å, β = 104.9°

Coordination sequences and vertex symbols:
   see Appendix A for a list of the coordination sequences and vertex symbols for the 12 T-atoms

Secondary building units: see *Compendium*

Composite building units:

d4r	cas	fer	non	ton

Materials with this framework type:
   *IM-12[1]
   ITQ-15[2]

IM-12   **Type Material Data**   **UTL**

Crystal chemical data: $[Ge_{13.8}Si_{62.2}O_{152}]$-**UTL**
monoclinic, $C2/m$

$$a = 29.8004\text{Å}, b = 13.9926\text{Å}, c = 12.3926\text{Å}, \beta = 105.185°^{(1)}$$

Framework density: $15.2 \text{ T}/1000\text{Å}^3$

Channels: [001] **14** 7.1 x 9.5* ↔ [010] **12** 5.5 x 8.5*

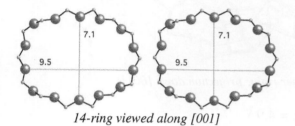 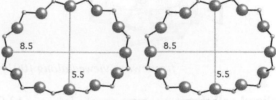

14-ring viewed along [001]   *12-ring viewed along [010]*

References:
(1) Paillaud, J.-L., Harbuzaru, B., Patarin, J. and Bats, N. *Science*, **304**, 990–992 (2004)
(2) Corma, A., Dîaz-Cabañas, M.J., Rey, F., Nicolopoulus, S. and Boulahya, K. *Chem. Commun.*, 1356–1357 (2004)

VET

Framework Type Data

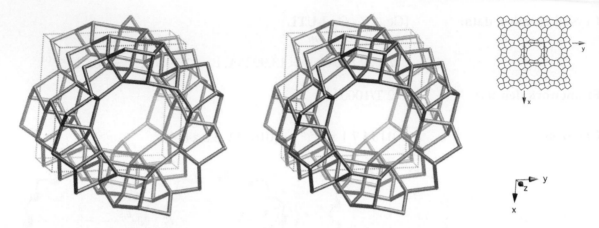

framework viewed along [001] (upper right: projection down [001])

Idealized cell data: tetragonal, $P\overline{4}$, $a = 13.0$Å, $c = 4.9$Å

Coordination sequences and vertex symbols:

T_1 (4,1)	4	12	24	39	61	93	133	179	209	246	$5 \cdot 5 \cdot 5 \cdot 6_2 \cdot 5 \cdot 7$
T_2 (4,1)	4	12	26	41	65	94	130	169	218	269	$5 \cdot 5 \cdot 5 \cdot 6 \cdot 7 \cdot 12_6$
T_3 (4,1)	4	12	24	41	65	95	128	169	218	270	$5 \cdot 6 \cdot 5 \cdot 6 \cdot 5 \cdot 6_2$
T_4 (4,1)	4	12	23	43	68	94	125	172	226	269	$5 \cdot 6 \cdot 5 \cdot 6 \cdot 5_2 \cdot 6$
T_5 (1,$\overline{4}$)	4	12	28	38	60	98	152	182	200	246	$5 \cdot 5 \cdot 5 \cdot 5 \cdot 8_2 \cdot 8_2$

Secondary building units: see *Compendium*

Composite building units:

cas

Materials with this framework type:
   *VPI-8[1]

VPI-8   **Type Material Data**   # VET

Crystal chemical data: $[Si_{17}O_{34}]$-**VET**
tetragonal, $P\overline{4}$, $a = 13.045$Å, $c = 5.034$Å [1]

Framework density: 19.8 T/1000Å^3

Channels: [001] **12** $5.9 \times 5.9^*$

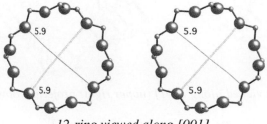

12-ring viewed along [001]

References:
(1) Freyhardt, C.C., Lobo, R.F., Khodabandeh, S., Lewis, J.E., Tsapatsis, M., Yoshikawa, M., Camblor, M.A., Pan, M., Helmkamp, M.M., Zones, S.I. and Davis, M.E. *J. Am. Chem. Soc.*, **118**, 7299–7310 (1996)

VFI

Framework Type Data

$P6_3/mcm$

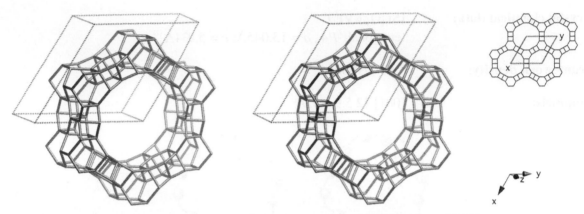

framework viewed along [001] (upper right: projection down [001])

Idealized cell data: hexagonal, $P6_3/mcm$, $a = 18.3$Å, $c = 8.6$Å

Coordination sequences and vertex symbols:

T_1 (24,1)	4	11	20	31	44	61	82	108	139	174	$4 \cdot 6_2 \cdot 6 \cdot 6_3 \cdot 6_2 \cdot 6_3$
T_2 (12,m)	4	10	18	30	44	60	80	106	135	168	$4 \cdot 6_3 \cdot 4 \cdot 6_3 \cdot 6 \cdot 6_4$

Secondary building units: 18 or 6 or 4-2

Composite building units:

dnc	afi	bog
double narsarsukite chain		

Materials with this framework type:

*VPI-5[1-3]

AlPO-54[2]

CoVPI-5[4]

FAPO-H1[5]

H1[6]

MCM-9[7]

TiVPI-5[8]

VPI-5   **Type Material Data**   **VFI**

Crystal chemical data: $|(H_2O)_{42}|$ $[Al_{18}P_{18}O_{72}]$-**VFI**
hexagonal, $P6_3$, $a = 18.975$Å, $c = 8.104$Å [3]

Framework density: 14.2 T/1000Å^3

Channels: [001] **18** 12.7 x 12.7*

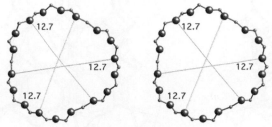

18-ring viewed along [001]

References:
(1)   Davis, M.E., Saldarriaga, C., Montes, C., Garces, J. and Crowder, C. *Nature*, **331**, 698–699 (1988)
(2)   Richardson Jr., J.W., Smith, J.V. and Pluth, J.J. *J. Phys. Chem.*, **93**, 8212–8219 (1989)
(3)   McCusker, L.B., Baerlocher, Ch., Jahn, E. and Bülow, M. *Zeolites*, **11**, 308–313 (1991)
(4)   Singh, P.S., Shaikh, R.A., Bandyopadhyay, R. and Rao, B.S. *Chem. Commun.*, 2255–2256 (1995)
(5)   Prasad, S. and Yang, T.C. *Catalysis Letters*, **28**, 269–275 (1994)
(6)   d'Yvoire, F. *Bull. Soc. Chim. France*, 1762–1776 (1961)
(7)   Derouane, E.G., Maistreiau, L., Gabelica, Z., Tuel, A., Nagy, J.B. and von Ballmoos, R. *Appl. Catal.*, **51**, L13-L20 (1989)
(8)   Tusar, N.N., Logar, N.Z., Arcon, I., ThibaultStarzyk, F. and Kaucic, V. *Croatica Chemica Acta*, **74**, 837–849 (2001)

VNI

Framework Type Data

P4₂/ncm

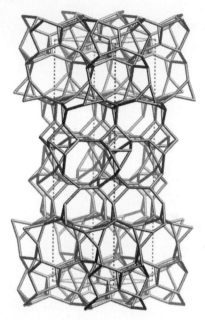
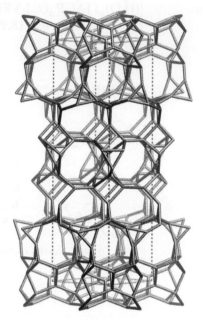

framework viewed along [110]

Idealized cell data: tetragonal, $P4_2/ncm$ (origin choice 2), $a = 10.0$Å, $c = 34.1$Å

Coordination sequences and vertex symbols:

T_1 (16,1)	4	11	23	39	63	93	126	170	210	255		$4 \cdot 8 \cdot 5 \cdot 8_2 \cdot 5_2 \cdot 8_2$
T_2 (8,m)	4	11	19	39	59	89	130	166	207	274		$4 \cdot 5_2 \cdot 5 \cdot 8 \cdot 5 \cdot 8$
T_3 (8,m)	4	9	20	37	61	92	117	152	201	246		$3 \cdot 4 \cdot 8 \cdot 8_2 \cdot 8 \cdot 8_2$
T_4 (8,m)	4	9	20	37	62	87	119	158	195	248		$3 \cdot 4 \cdot 8_2 \cdot 8_3 \cdot 8_2 \cdot 8_3$
T_5 (8,m)	4	9	21	41	59	85	133	155	195	261		$3 \cdot 4 \cdot 8 \cdot 8_3 \cdot 8 \cdot 8_3$
T_6 (8,m)	4	10	18	39	65	83	119	169	218	236		$3 \cdot 8_3 \cdot 5 \cdot 5_2 \cdot 5 \cdot 5_2$
T_7 (4,2mm)	4	10	18	36	64	82	118	176	202	264		$3 \cdot 8_2 \cdot 5 \cdot 5 \cdot 5 \cdot 5$

Secondary building units: see *Compendium*

Composite building units:

vsv

Materials with this framework type:

*VPI-9[1]

VPI-9   Type Material Data   **VNI**

Crystal chemical data: |Rb$_{44}$K$_4$ (H$_2$O)$_{48}$| [Zn$_{24}$Si$_{96}$O$_{240}$]-**VNI**
tetragonal, $P4_12_12$, $a = 9.884$Å, $c = 73.650$Å $^{(1)}$
(Relationship to unit cell of Framework Type: $a' = a$, $b' = b$, $c' = 2c$)

Framework density: 16.7 T/1000Å^3

Channels: {<110> **8** 3.1 x 4.0 ↔ [001] **8** 3.5 x .3.6}***

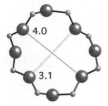 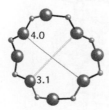 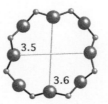 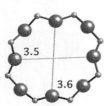

8-ring along <110>   *8-ring viewed along [001]*

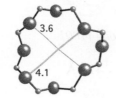 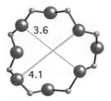

2nd 8-ring along <110>

References:
(1) McCusker, L.B., Grosse-Kunstleve, R.W., Baerlocher, Ch., Yoshikawa, M. and Davis, M.E. *Microporous Materials*, **6**, 295–309 (1996)

VSV

Framework Type Data

I4₁/amd

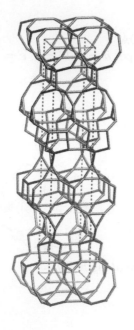
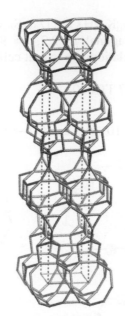

framework viewed along [100]

Idealized cell data: tetragonal, *I4₁/amd*, $a = 7.2$Å, $c = 41.8$Å

Coordination sequences and vertex symbols:

T₁ (16,*m*)	4	9	21	42	61	81	123	159	198	246		$3·4·8_2·9_4·8_2·9_4$
T₂ (16,*m*)	4	11	21	40	61	93	122	151	195	251		$4·5_2·5·8·5·8$
T₃ (4, $\overline{4}m2$)	4	8	20	48	56	84	120	160	212	240		$3·3·9_4·9_4·9_4·9_4$

Secondary building units: see *Compendium*

Composite building units:

     lov   *vsv*

Materials with this framework type:
    *VPI-7[1,2]
    Gaultite[3]
    VSV-7#[4]

VPI-7   Type Material Data   **VSV**

Crystal chemical data: $|Na_{26}H_6 (H_2O)_{44}|$ $[Zn_{16}Si_{56}O_{144}]$-**VSV**
orthorhombic, $Fdd2$, $a = 39.88$Å, $b = 10.326$Å, $c = 10.219$Å [2]
(Relationship to unit cell of Framework Type:

$a' = c$, $b' = a\sqrt{2}$, $c' = b\sqrt{2}$
or, as vectors, $\mathbf{a'} = \mathbf{c}$, $\mathbf{b'} = \mathbf{b} - \mathbf{a}$, $\mathbf{c'} = \mathbf{a} + \mathbf{b}$)

Framework density: 17.1 T/1000Å^3

Channels: $[01\bar{1}]$ **9** 3.3 x 4.3* ↔ [011] **9** 2.9 x 4.2* ↔ [011] **8** 2.1 x 2.7*

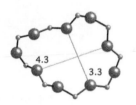 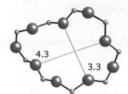 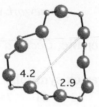 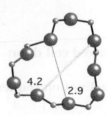

9-ring along $[01\bar{1}]$ 9-ring along [011]

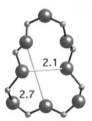 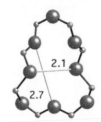 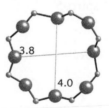 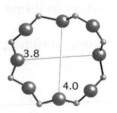

8-ring along [011]   8-ring viewed along [100]

References:
(1) Annen, M.J., Davis, M.E., Higgins, J.B. and Schlenker, J.L. *Chem. Commun.*, 1175–1176 (1991)
(2) Röhrig, C., Gies, H. and Marler, B. *Zeolites*, **14**, 498–503 (1994)
(3) Ercit, T.S. and van Velthuizen, J. *Can. Mineral.*, **32**, 855–863 (1994)
(4) Röhrig, C., Dierdorf, I. and Gies, H. *J. Phys. Chem. Solids*, **56**, 1369–1376 (1995)

WEI

Framework Type Data

Cccm

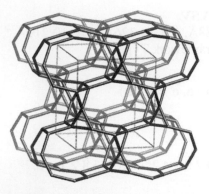

framework viewed along [001]

Idealized cell data: orthorhombic, *Cccm*, $a = 11.8$Å, $b = 10.3$Å, $c = 10.0$Å

Coordination sequences and vertex symbols:

T_1 (16,1)	4	9	18	32	51	74	98	126	163	199
T_2 (4,222)	4	8	18	32	52	70	98	132	152	200

3·4·6·8·8·10
3·3·6·6·10·10

Secondary building units: spiro-5

Composite building units:

lov vsv

Materials with this framework type:
  *Weinebeneite[1]

Weinebeneite

Crystal chemical data: |Ca$_4$ (H$_2$O)$_{16}$| [Be$_{12}$P$_8$ O$_{32}$ (OH)$_8$]-**WEI**
monoclinic, *Cc*

$a = 11.897$Å, $b = 9.707$Å, $c = 9.633$Å, $\beta = 95.76°$ [(1)]

Framework density: 18.1 T/1000Å^3

Channels: [001] **10** 3.1 x 5.4* $\leftrightarrow$ [100] **8** 3.3 x 5.0*

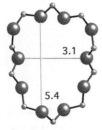 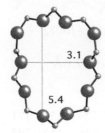 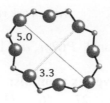 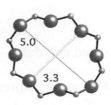

10-ring viewed along [001]   *8-ring viewed along [100]*

References:
(1) Walter, F. *Eur. J. Mineral.*, **4**, 1275–1283 (1992)

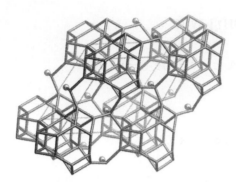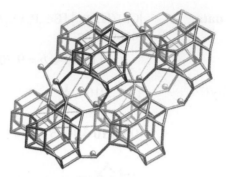

framework viewed along [001]

Idealized cell data: hexagonal, $P\bar{6}2m$, $a = 13.6$Å, $c = 7.6$Å

Coordination sequences and vertex symbols:

T_1 (12,1)	4	9	16	27	46	73	102	129	157	191	$4 \cdot 4 \cdot 4 \cdot 6 \cdot 6 \cdot 10_2$
T_2 (6,m)	4	9	19	34	49	67	94	125	157	195	$4 \cdot 8 \cdot 4 \cdot 8 \cdot 6 \cdot 8_2$
T_3 (2,-6)	3	9	21	36	53	69	90	119	156	201	$8_2 \cdot 8_2 \cdot 8_2$

Secondary building units: see *Compendium*

Composite building units:

        dsc   *d6r*   *can*

double sawtooth
chain

Materials with this framework type:
  *Wenkite[1,2]

Wenkite | **Type Material Data** | **-WEN**

Crystal chemical data: $|Ba_4 (Ca,Na_2)_3 H_2O (SO_4)_3| [Al_8Si_{12}O_{39} (OH)_2]$- **-WEN**
hexagonal, $P\bar{6}2m$, $a = 13.511Å$, $c = 7.462Å^{(2)}$

Framework density: 17 T/1000Å³

Channels: <100> **10** 2.5 x 4.8** ↔ [001] **8** 2.3 x 2.7*

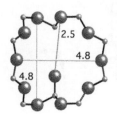
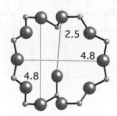
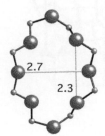
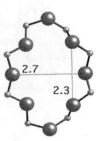

10-ring viewed along <100>   *8-ring viewed along [001]*

References:
(1) Wenk, H.-R. *Z. Kristallogr.*, **137**, 113–126 (1973)
(2) Merlino, S. *Acta Crystallogr.*, **B30**, 1262–1266 (1974)

YUG

Framework Type Data

C2/m

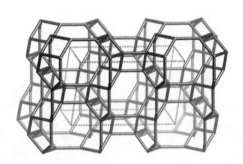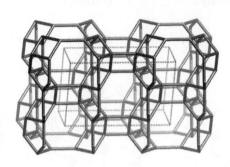

framework viewed along [001]

Idealized cell data: monoclinic, *C2/m*, *a* = 10.2Å, *b* = 13.8Å, *c* = 6.8Å, β = 111.5°

Coordination sequences and vertex symbols:

T$_1$ (8,1)	4	11	22	39	61	88	120	155	192	241
T$_2$ (8,1)	4	10	22	39	61	89	118	153	198	241

4·5·5·5·8·8

4·4·5·8$_2$·5·8$_2$

Secondary building units: 8 or 4

Materials with this framework type:
   *Yugawaralite[1-3]
   Sr-Q[4]
   Yugawaralite, Hvalfjördur, Iceland[5]

Yugawaralite Type Material Data **YUG**

Crystal chemical data: |Ca$_2$ (H$_2$O)$_8$| [Al$_4$Si$_{12}$O$_{32}$]-**YUG**

monoclinic, *Pc*, *a* = 6.73Å, *b* = 13.95Å, *c* = 10.03Å, β = 111.5°$^{(2)}$
(Relationship to unit cell of Framework Type: *a'* = *c*, *b'* = *b*, *c'* = *a*)

Framework density: 18.3 T/1000Å^3

Channels: [100] **8** 2.8 x 3.6* ↔ [001] **8** 3.1 x 5.0*

 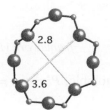 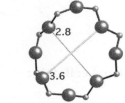

8-ring viewed along [100]   *8-ring viewed along [001]*

References:
(1) Kerr, I.S. and Williams, D.J. *Z. Kristallogr.*, **125**, 220–225 (1967)
(2) Kerr, I.S. and Williams, D.J. *Acta Crystallogr.*, **B25**, 1183–1190 (1969)
(3) Leimer, H.W. and Slaughter, M. *Z. Kristallogr.*, **130**, 88–111 (1969)
(4) Hawkins, D.B. *Mater. Res. Bull.*, **2**, 951–958 (1967)
(5) Kvick, A., Artioli, G. and Smith, J.V. *Z. Kristallogr.*, **174**, 265–281 (1986)

ZON

Framework Type Data

Pbcm

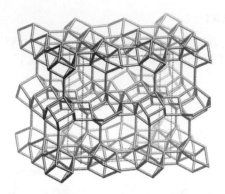
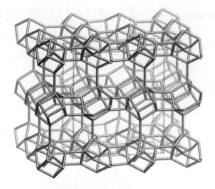

framework viewed along [100]

Idealized cell data: orthorhombic, *Pbcm*, $a = 6.9$Å, $b = 14.9$Å, $c = 17.2$Å

Coordination sequences and vertex symbols:

T_1 (8,1)	4	10	21	34	47	72	108	136	162	200	$4·8·4·8_2·6_3·8$
T_2 (8,1)	4	9	19	33	53	78	100	126	166	213	$4·4·4·8·6·6_3$
T_3 (8,1)	4	10	18	33	57	77	95	129	172	209	$4·6·4·6·6·8$
T_4 (8,1)	4	9	17	32	53	74	98	128	165	208	$4·6·4·6_2·4·8$

Secondary building units: 6-2 or 4-4- or 4

Composite building units:
   sti

Materials with this framework type:
   *ZAPO-M1[1]
   GaPO-DAB-2[2,3]
   UiO-7[4,5]

ZAPO-M1   Type Material Data   **ZON**

Crystal chemical data: $|(C_4H_{12}N)_8|$ $[Zn_8Al_{24}P_{32}O_{128}]$-**ZON**
         $C_4H_{12}N$ = tetramethylammonium
    orthorhombic, *Pbca*, $a = 14.226$Å, $b = 15.117$Å, $c = 17.557$Å $^{(1)}$
    (Relationship to unit cell of Framework Type: $a' = 2a$, $b' = b$, $c' = c$)

Framework density: 17 T/1000Å^3

Channels: [100] **8** 2.5 x 5.1* ↔ [010] **8** 3.7 x 4.4*

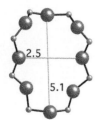 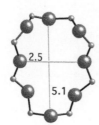 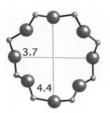 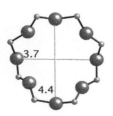

       8-ring viewed along [100]   *8-ring viewed along [010]*

References:
(1) Marler, B., Patarin, J. and Sierra, L. *Microporous Materials*, **5**, 151–159 (1995)
(2) Schott-Darie, C., Kessler, H., Soulard, M., Gramlich, V. and Benazzi, E. *Stud. Surf. Sci. Catal.*, **84**, 101–108 (1994)
(3) Meden, A., Grosse-Kunstleve, R.W., Baerlocher, Ch. and McCusker, L.B. *Z. Kristallogr.*, **212**, 801–807 (1997)
(4) Akporiaye, D.E., Fjellvag, H., Halvorsen, E.N., Hustveit, J., Karlsson, A. and Lillerud, K.P. *Chem. Commun.*, 601–602 (1996)
(5) Akporiaye, D.E., Fjellvag, H., Halvorsen, E.N., Hustveit, J., Karlsson, A. and Lillerud, K.P. *J. Phys. Chem.*, **100**, 16641–16646 (1996)

APPENDIX A

CHA

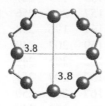
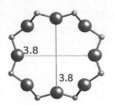

8-ring viewed normal to [001]

GIS

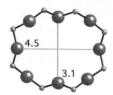
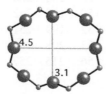
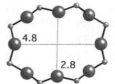
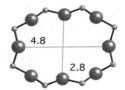

8-ring viewed along [100]   *8-ring viewed along [010]*

IMF

Coordination sequences and vertex symbols:

T_1 (8,1)	4	9	18	33	56	84	108	135	179	227	276	326	$4·5·4·6·4·10$
T_1 (16,1)	4	12	21	39	65	92	119	157	208	256	290	357	$5·5·5·5_2·5·10_3$
T_2 (16,1)	4	12	22	37	58	94	129	161	196	248	317	374	$5·5·5·6·5·6_2$
T_3 (16,1)	4	12	21	40	60	88	121	159	202	249	296	358	$5·5·5·5_2·5·10_2$
T_4 (16,1)	4	12	23	39	61	87	122	162	201	242	301	364	$5·5_2·5·6·5·10_3$
T_5 (16,1)	4	12	24	39	62	92	126	159	207	251	305	362	$5·5_2·5·6·5·10_3$
T_6 (16,1)	4	12	24	40	60	91	126	167	204	250	307	368	$5·5·5·10_3·5_2·6$
T_7 (16,1)	4	12	21	38	60	89	123	161	202	249	305	362	$5·5·5·5·5·6$
T_8 (16,1)	4	12	23	38	60	90	126	162	205	256	309	367	$5·6_2·5·10_2·5_2·6$
T_9 (16,1)	4	12	25	39	57	88	128	160	195	246	309	369	$5·6·5·6_2·5·10_2$
T_{10} (16,1)	4	12	21	41	64	89	125	161	209	253	303	360	$5·5·5·5·5_2·10_2$
T_{11} (16,1)	4	11	21	37	61	87	121	159	201	251	303	362	$4·5_2·5·6_2·10·10$
T_{12} (16,1)	4	11	23	39	62	93	131	167	201	248	312	377	$4·5·5·6_2·6·10_2$
T_{13} (8,m..)	4	12	20	38	63	94	118	150	206	262	291	347	$5·5_2·5·5_2·10_2·0$
T_{14} (8,m..)	4	12	22	34	63	92	119	160	201	243	300	360	$5·5·5·5·5_2·10_2$
T_{15} (8,m..)	4	12	19	39	62	88	123	156	199	256	302	358	$5·5_2·5·5_2·6·*$
T_{16} (8,m..)	4	12	21	35	62	90	123	161	202	252	306	360	$5·5·5·5·5_2·6$
T_{17} (8,m..)	4	12	22	35	57	90	126	159	197	241	301	370	$5_2·6_2·5_2·6_2·10·12_2$
T_{18} (8,m..)	4	12	21	33	55	94	136	157	185	244	315	384	$5_2·6_2·5_2·6_2·6·*$
T_{19} (8,m..)	4	10	21	37	61	94	126	162	200	249	312	368	$4·5·4·5·6·10_2$
T_{20} (8,m..)	4	11	21	35	61	97	123	153	205	260	313	363	$4·6·5·5·5·5$
T_{21} (8,m..)	4	11	20	33	59	93	124	150	189	252	312	366	$4·6_2·5·5·5·5$
T_{22} (8,m..)	4	11	20	36	62	92	119	152	199	251	305	355	$4·6_2·5·5·5·5$
T_{23} (8,m..)	4	11	22	37	66	97	120	153	208	267	297	354	$4·10_3·5·5·5·5$
T_{24} (8,m..)	4	10	22	37	61	89	112	156	215	257	294	348	$4·5·4·5·10·10_2$

IWW

Coordination sequences and vertex symbols:

T_1 (8,1)	4	9	18	33	56	84	108	135	179	227	276	326	$4·5·4·6·4·10$
T_2 (8,1)	4	9	18	33	55	84	114	144	177	220	276	330	$4·6·4·6_2·4·10_4$
T_3 (8,1)	4	9	19	35	55	80	110	142	174	216	275	324	$4·6·4·6_2·4·10_2$
T_4 (8,1)	4	9	18	34	56	79	104	138	177	219	267	320	$4·5·4·6·4·10$
T_5 (8,1)	4	12	21	36	53	76	106	143	179	224	271	325	$5·6·5·6_2·5·8$
T_6 (8,1)	4	12	18	33	55	77	105	138	178	226	265	317	$5·5·5·6·5_2·10$
T_7 (8,1)	4	12	20	36	55	75	106	138	176	219	265	327	$5·5_2·8·6·8_2$
T_8 (8,1)	4	11	18	33	53	78	116	146	180	219	272	323	$4·6_2·5·6·5·10_2$
T_9 (8,1)	4	11	24	38	52	77	108	143	181	211	271	342	$4·8_2·5·5·5·8_2$
T_{10} (8,1)	4	11	20	35	54	75	107	143	172	216	275	319	$4·5_2·5·5·5·8_2$
T_{11} (8,1)	4	11	21	34	56	76	106	145	190	225	265	320	$4·5·5·6_2·5·12_7$
T_{12} (8,1)	4	10	20	34	52	78	110	156	191	216	261	331	$4·4·5·6_2·5·12_7$
T_{13} (4,..m)	4	11	20	29	50	82	104	129	180	220	277	321	$4·5_2·5·6·5·6$
T_{14} (4,..m)	4	11	20	29	48	85	111	134	165	232	282	312	$4·5_2·6·6_1·6·6_9$
T_{15} (4,..m)	4	11	22	28	50	80	110	126	174	217	268	326	$4·5_2·6·6_2·6·6_2$
T_{16} (4,..m)	4	11	20	31	46	79	105	131	163	225	267	313	$4·5_2·5·6·5·6$

MFI

Coordination sequences and vertex symbols:

T_1 (8,1)	4	12	22	41	61	88	125	159	198	250	$5 \cdot 5 \cdot 5 \cdot 10_2 \cdot 5_2 \cdot 6$
T_2 (8,1)	4	12	22	39	64	91	117	158	209	247	$5 \cdot 5 \cdot 5 \cdot 5_2 \cdot 5 \cdot 10_3$
T_3 (8,1)	4	12	23	37	62	91	120	157	206	250	$5 \cdot 5 \cdot 5 \cdot 5_2 \cdot 5 \cdot 10_3$
T_4 (8,1)	4	12	21	36	61	90	122	159	196	251	$5 \cdot 5 \cdot 5 \cdot 5_2 \cdot 5 \cdot 6$
T_5 (8,1)	4	12	24	38	63	93	123	157	206	247	$5 \cdot 5 \cdot 5 \cdot 6_2 \cdot 5 \cdot 10_3$
T_6 (8,1)	4	12	22	40	61	88	124	156	197	253	$5 \cdot 5 \cdot 5 \cdot 5 \cdot 5_2 \cdot 10_3$
T_7 (8,1)	4	12	24	38	56	90	132	164	193	241	$5 \cdot 6_2 \cdot 5_2 \cdot 6_2 \cdot 10 \cdot 10$
T_8 (8,1)	4	12	21	37	63	90	121	155	201	253	$5 \cdot 5 \cdot 5 \cdot 5 \cdot 5 \cdot 6$
T_9 (8,1)	4	11	23	39	62	93	119	153	204	254	$4 \cdot 5 \cdot 5 \cdot 6_2 \cdot 5 \cdot 10_3$
T_{10} (8,1)	4	11	22	36	61	93	120	154	200	255	$4 \cdot 5 \cdot 5 \cdot 6_2 \cdot 5 \cdot 10_2$
T_{11} (8,1)	4	12	22	38	59	92	125	159	202	250	$5 \cdot 5 \cdot 5 \cdot 6 \cdot 5 \cdot 6_2$
T_{12} (8,1)	4	12	23	38	59	89	126	161	196	246	$5 \cdot 6_2 \cdot 5 \cdot 10_2 \cdot 5_2 \cdot 6_2$

MWW

Coordination sequences and vertex symbols:

T_1 (12,.m.)	4	10	20	32	52	76	111	146	185	225	$4_2 \cdot 6 \cdot 5 \cdot 5 \cdot 5 \cdot 5$
T_2 (12,.m.)	4	12	22	35	51	81	109	137	175	218	$5 \cdot 5 \cdot 5 \cdot 5 \cdot 6 \cdot 10_2$
T_3 (12,.m.)	4	10	21	40	62	78	95	128	177	228	$4 \cdot 5 \cdot 4 \cdot 5 \cdot 6 \cdot 10_2$
T_4 (12,..m)	4	11	18	32	52	78	107	147	187	215	$4 \cdot 10_2 \cdot 5_2 \cdot 6_2 \cdot 5_2 \cdot 6_2$
T_5 (12,..m)	4	11	22	32	53	79	113	144	176	220	$4 \cdot 5 \cdot 5 \cdot 6_2 \cdot 5 \cdot 6_2$
T_6 (4,3m)	4	10	20	34	54	87	114	139	188	244	$4 \cdot 10_4 \cdot 4 \cdot 10_4 \cdot 4 \cdot 10_4$
T_7 (4,3m)	4	10	16	30	49	77	100	138	181	214	$4 \cdot 6_2 \cdot 4 \cdot 6_2 \cdot 4 \cdot 6_2$
T_8 (4,3m)	4	12	22	31	52	74	112	142	166	204	$5 \cdot 6_2 \cdot 5 \cdot 6_2 \cdot 5 \cdot 6_2$

PAU

Coordination sequences and vertex symbols:

T_1 (96,1)	4	9	18	31	47	68	91	117	151	188	$4 \cdot 4 \cdot 4 \cdot 8_2 \cdot 8 \cdot 8$
T_2 (96,1)	4	9	18	32	49	70	95	122	154	191	$4 \cdot 4 \cdot 4 \cdot 8_2 \cdot 8 \cdot 8$
T_3 (96,1)	4	9	18	32	49	69	94	123	153	186	$4 \cdot 4 \cdot 4 \cdot 8_2 \cdot 8 \cdot 8$
T_4 (96,1)	4	9	18	32	48	67	92	121	152	185	$4 \cdot 4 \cdot 4 \cdot 8_2 \cdot 8 \cdot 8$
T_5 (96,1)	4	9	18	31	47	68	92	119	152	188	$4 \cdot 4 \cdot 4 \cdot 8_2 \cdot 8 \cdot 8$
T_6 (96,1)	4	9	17	29	45	65	89	117	149	185	$4 \cdot 4 \cdot 4 \cdot 6 \cdot 8 \cdot 8$
T_7 (48,2)	4	9	17	30	47	66	88	113	144	183	$4 \cdot 4 \cdot 4 \cdot 6 \cdot 8 \cdot 8$
T_8 (48,2)	4	9	18	32	49	69	95	123	152	188	$4 \cdot 4 \cdot 4 \cdot 8_2 \cdot 8 \cdot 8$

TUN

Coordination sequences and vertex symbols:

T_1 (8,1)	4	12	22	34	59	88	116	157	198	235	285	347	$5·5·5·5·5_2·10$
T_2 (8,1)	4	12	20	37	60	88	117	147	192	246	299	347	$5·5·5·5_2·5·6$
T_3 (8,1)	4	12	21	39	62	84	115	155	201	242	284	338	$5·5·5·5·5_2·10_3$
T_4 (8,1)	4	12	22	37	58	89	124	158	190	234	298	363	$5·5·5·6·5·6_2$
T_5 (8,1)	4	12	24	38	59	88	119	156	195	240	292	353	$5·5·5·10_3·5_2·6$
T_6 (8,1)	4	12	23	38	59	83	118	159	197	239	285	346	$5·6·5·10_2·5_2·6_2$
T_7 (8,1)	4	12	21	37	56	90	121	154	198	234	292	350	$5·5·5·5·5·6$
T_8 (8,1)	4	12	22	36	55	87	129	158	186	233	291	355	$5_2·6_2·5_2·6_2·10·*$
T_9 (8,1)	4	12	22	33	55	88	124	153	187	238	292	350	$5·6_2·5·12_2·5_2·6$
T_{10} (8,1)	4	12	20	36	60	87	113	154	196	244	289	336	$5·5·5·5_2·5·10$
T_{11} (8,1)	4	12	19	37	62	87	115	153	194	243	291	345	$5·5·5·5·5_2·6$
T_{12} (8,1)	4	12	20	41	61	84	119	146	197	250	285	338	$5·5_2·5·5_2·10_2·*$
T_{13} (8,1)	4	12	24	38	61	91	121	154	194	244	297	346	$5·5_2·5·6·5·10_3$
T_{14} (8,1)	4	12	22	37	57	86	121	160	194	236	288	351	$5·5·5·6·5_2·10$
T_{15} (8,1)	4	12	21	36	66	92	112	151	202	250	283	334	$5·5·5·52·5·10_3$
T_{16} (8,1)	4	11	21	34	57	90	123	150	187	238	296	361	$4·5_2·5·6_2·6·12_2$
T_{17} (8,1)	4	11	24	39	58	86	122	158	190	237	301	359	$4·5·5·6_2·6·10_2$
T_{18} (8,1)	4	11	22	36	55	87	123	159	188	237	295	353	$4·6_2·5·5·5·8_2$
T_{19} (8,1)	4	11	21	35	58	86	117	150	198	241	288	352	$4·5·5·5·5·8_2$
T_{20} (8,1)	4	11	21	36	60	87	113	150	199	244	289	344	$4·5_2·5·6·10·10$
T_{21} (8,1)	4	11	23	37	60	88	121	155	200	235	292	352	$4·6_2·5·5·5·10$
T_{22} (8,1)	4	11	21	38	60	86	116	152	197	247	291	339	$4·5_2·5·5·5·10$
T_{23} (8,1)	4	11	23	38	56	87	122	152	193	241	296	351	$4·5·5·6·10·10$
T_{24} (8,1)	4	10	22	39	62	83	109	156	201	241	293	348	$4·5·4·5·10·10_2$

UTL

Coordination sequences and vertex symbols:

T_1 (8,1)	4	9	18	32	53	80	104	129	171	217	264	308	$4·5·4·5·4·12$
T_2 (8,1)	4	9	18	32	53	79	105	130	166	220	263	311	$4·5·4·5·4·12$
T_3 (8,1)	4	12	22	34	49	73	102	144	181	213	246	306	$5·5·5·6·52·12$
T_4 (8,1)	4	12	21	34	49	71	102	139	183	215	251	298	$5·5·5·5_2·5·12$
T_5 (8,1)	4	12	21	34	48	73	106	140	176	208	255	310	$5·5·5·5·5_2·6$
T_6 (8,1)	4	12	22	33	52	76	107	144	173	208	259	311	$5·5·5·6_2·5·12$
T_7 (8,1)	4	12	20	31	53	76	104	140	170	212	255	315	$5·5·5·5_2·5·12$
T_8 (4,m)	4	12	24	34	46	71	107	147	176	215	249	300	$5·5·5·5·14_6·*$
T_9 (4,m)	4	12	20	34	48	68	107	141	178	211	242	298	$5·5_2·5·5_2·14_2·*$
T_{10} (4,m)	4	12	20	28	51	73	100	144	172	208	256	287	$5·5·5·5·5·5_2$
T_{11} (4,m)	4	12	20	36	50	67	102	145	178	223	252	284	$5·5_2·5·5_2·14_2·*$
T_{12} (4,m)	4	12	22	30	49	73	106	145	179	199	250	315	$5·5·5·5·5·6_2$

APPENDIX B

Rules for Framework Type Assignment

The following is a set of rules to be applied by the Structure Commission of the IZA in assigning a three-letter code to a new framework type. The materials of interest are generally defined as open 4-connected 3D nets which have the general (approximate) composition AB_2, where A is a tetrahedrally connected atom and B is any 2-connected atom, which may or may not be shared, between two neighboring A atoms. Inclusion of other microporous materials is left to the discretion of the IZA Structure Commission, depending on the interest of the molecular sieve science community at large.

RULES

(i) The IZA Structure Commission is the only body that can coordinate the assignment of a code to such frameworks. The framework types are idealized, with no reference to actual materials, symmetry, composition, etc. and, therefore, ONLY refer to the connectivity.

(ii) The code for a 4-connected 3D framework type shall consist of three capital letters. Other frameworks of interest shall be indicated by a (-) in front of the code.

(iii) The three letters of the code shall be mnemonic and must refer to an actual material (i.e. the type material). These materials are chosen to be:

- mineral names (rules of the International Mineralogical Association followed).

- commonly accepted synthetic material types.

- in the absence of the above, the workers who first determined the structure have priority in assigning the name.

(iv) No mnemonic code can be assigned without the structure being determined with the following exception:

For "polytypic" materials, codes can be assigned as useful. Such codes shall be marked with an asterisk.

(v) Codes of framework types which turn out to be in error are discredited. Later use of the code is not permitted.

(vi) For all cases where a decision of the Structure Commission is required, a two thirds majority vote of the full commission shall be required. Such votes are taken verbally at a meeting of the Structure Commission or are done in writing on the initiative of the Chairperson and Co-chairperson. All evidence substantiating a new framework type must accompany the ballot. The members are obliged to respond to this request within one month and the Chairperson and Co-chairperson will make all possible effort to solicit replies from all members.

APPENDIX C

Secondary Building Units and their Symbols[a]

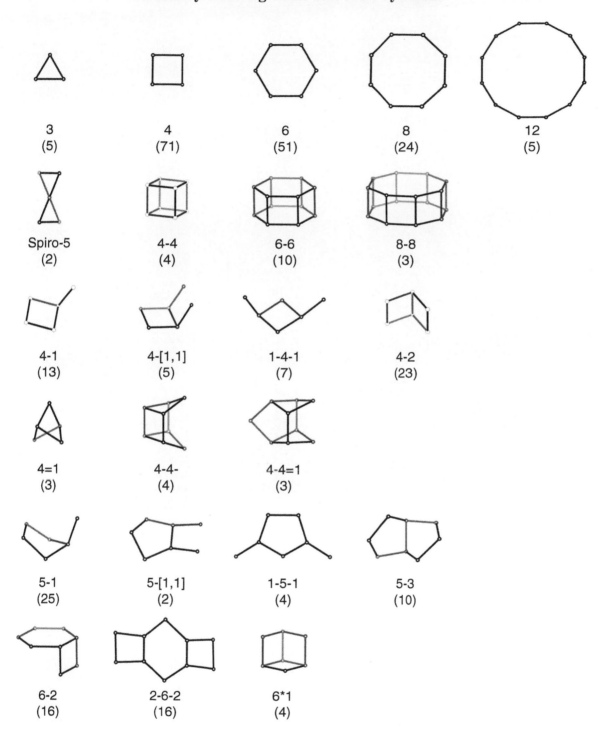

3
(5)

4
(71)

6
(51)

8
(24)

12
(5)

Spiro-5
(2)

4-4
(4)

6-6
(10)

8-8
(3)

4-1
(13)

4-[1,1]
(5)

1-4-1
(7)

4-2
(23)

4=1
(3)

4-4-
(4)

4-4=1
(3)

5-1
(25)

5-[1,1]
(2)

1-5-1
(4)

5-3
(10)

6-2
(16)

2-6-2
(16)

6*1
(4)

[a] Number in parenthesis = frequency of occurrence

APPENDIX D

Selected Composite Building Units

ordered by number of T-atoms in the unit

Each unit is identified with a lower-case three-character code in italics and the number of T-atoms in the unit. Framework types containing the unit are listed below each unit.

lov 5T	*nat* 6T	*vsv* 6T	*mei* 7T
LOV, NAB, OBW, OSO, RSN, VSV, WEI	**EDI, NAT, THO**	**LOV, NAB, OBW, -RON, RSN, VNI, VSV, WEI**	**AFN, CGF, MEI, USI** (see also *d4r - 8T*, *sti - 8T* and *bph - 14T*)
d4r 8T	*mor* 8T	*sti* 8T	*bea* 10T
ACO, AFY, AST, ASV, BEC, -CLO, DFO, ISV, ITH, ITW, IWR, IWV, IWW, LTA, UFI, UOZ, UTL	***BEA, BEC, DAC, EON, EPI, IMF, ISV, IWW, MEL, MFI, MOR, MSE, RWR, TUN** (see also *fer - 13T*)	**AFR, DFO, OWE, SAO, SBE, SBS, SBT, SFO, STI, ZON** (see also *d4r - 8T*)	***BEA, CON, IFR, MSE, STT**
bre 10T	*jbw* 10T	*mtt* 11T	*afi* 12T
BOG, BRE, CON, HEU, IWR, IWW, RRO, STI, TER	**JBW, MTT, MTW, SFE, SFN, SSY, TON**	**CFI, IMF, MFS, MTT, SFE, SSY, SZR, TON, TUN** (see also *non - 15T*)	**AEL, AET, AFI, AFO, ATV, DON, SFH, VFI** (see also *bph - 14T*)

afs 12T	ats 12T	bog 12T	cas 12T
AFS, BPH (see also *aww - 16T*)	**ATS, IMF, OSI**	**AEL, AET, AFI, AFO, AHT, ATV, BOG, CGF, DFO, LAU, TER, USI, VFI**	**BOG, CAS, CFI, EUO, GON, IHW, IWV, MFI, MTW, NES, NSI, SFF, SFH, SFN, STF, STT, TER, TUN, UTL, VET**

d6r 12T	lau 12T	rth 12T	stf 12T
AEI, AFT, AFX, CHA, EAB, EMT, ERI, FAU, GME, KFI, LEV, LTL, LTN, MOZ, MSO, MWW, OFF, SAS, SAT, SAV, SBS, SBT, SZR, TSC, -WEN	**ASV, ATO, BCT, CON, DFO, EZT, IFR, ITH, IWR, IWW, LAU, MSO, OSI, -RON, SAO, TUN, UOZ** (see also *mtw - 14T*)	**ITE, RTH, UFI**	**SFF, STF, TUN, IMF, IWW**

bik 13T	fer 13T	abw 14T	bph 14T
BIK, CAS, MTT, MTW, NSI, SFE, TON	**CDO, FER, IMF, MFS, UTL**	**ABW, ATT, JBW**	**AFS, BPH, EZT**

mel 14T	mfi 14T	mtw 14T	non 15T
CON, DON, ITH, IWR, IWW, MEL, MFI, MWW, SFG	**MEL, MFI, MTF**	***BEA, BEC, GON, ISV, MSE, MTW, SFH, SFN**	**EUO, IHW, IWV, NES, NON, UTL**

ton 15T	aww 16T	d8r 16T	rte 16T
CFI, IHW, IWV, MTT, NES, SFE, SSY, TON, UTL	AWW, SAO	MER, PAU, RHO, SBE, TSC	RTE, RUT

can 18T	mso 18T	gis 20T	mtn 20T
AFG, CAN, ERI, FAR, FRA, GIU, LIO, LOS, LTL, LTN, MAR, MOZ, OFF, SAT, SBS, SBT, TOL, -WEN	MSO, SZR	ATT, GIS, SIV	DDR, DOH, MEP, MTN

atn 24T	gme 24T	sod 24T	los 30T
ATN, SBE	AFT, AFX, EAB, EON, GME, MAZ, OFF	EMT, FAR, FAU, FRA, GIU, LTA, LTN, MAR, SOD, TSC	FRA, LIO, LOS, TOL

clo 32T	pau 32T	cha 36T	lio 42T
AWW, -CLO	KFI, MER, MOZ, PAU	AFT, CHA	AFG, FAR, LIO, MAR, TOL

378

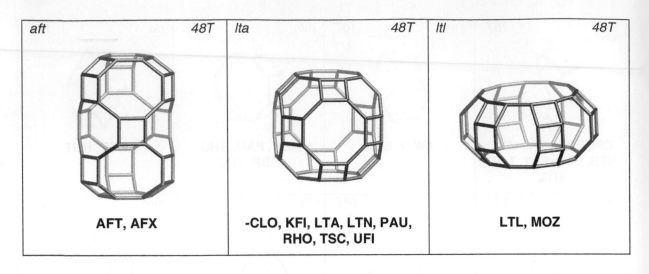

aft	48T	lta	48T	ltl	48T
AFT, AFX		**-CLO, KFI, LTA, LTN, PAU, RHO, TSC, UFI**		**LTL, MOZ**	

Selected Chains

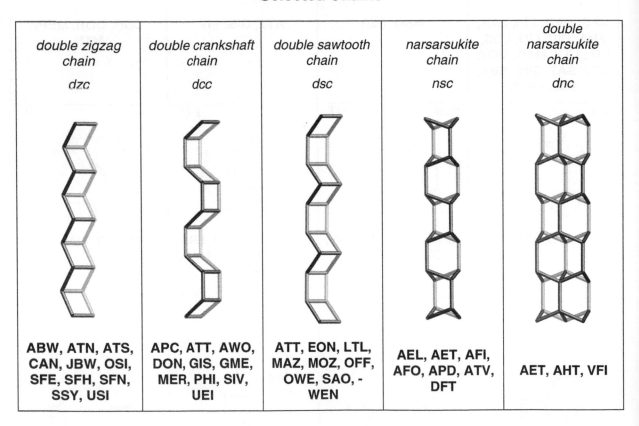

double zigzag chain	*double crankshaft chain*	*double sawtooth chain*	*narsarsukite chain*	*double narsarsukite chain*
dzc	*dcc*	*dsc*	*nsc*	*dnc*
ABW, ATN, ATS, CAN, JBW, OSI, SFE, SFH, SFN, SSY, USI	APC, ATT, AWO, DON, GIS, GME, MER, PHI, SIV, UEI	ATT, EON, LTL, MAZ, MOZ, OFF, OWE, SAO, -WEN	AEL, AET, AFI, AFO, APD, ATV, DFT	AET, AHT, VFI

Composite Building Unit Nomenclature Comparison

Atlas	T-atoms	Compendium[a]		J.V. Smith[b]		Occurrence
abw	14	-	kdq	$4^2 6^2 8^2$-b	*mm*2	3
afi	12	-	afi	$6^3 6^2$	$\bar{6}2m$	8
afs	12	-	afs	$4^2 4^2 6^2 8^1$	*mm*2	2
aft	48	{2 [$4^{15}6^2 8^9$]}	aft	$4^6 4^6 4^3 6^2 8^6 8^3$	$\bar{6}2m$	2
atn	24	{1 [$4^8 6^4 8^2$]}	ocn	$4^8 6^4 8^2$	4/*mmm*	2
ats	12	[$4^2 6^4$]-c	oth	$4^2 6^2 6^2$	*mm*2	2
aww	16	[$4^6 6^4$]	aww	$4^4 4^2 6^4$	$\bar{4}2m$	2
bea	10	[$4^3 5^4$]	wwt	$4^2 4^1 5^2 6^1$	*mm*2	5
bog	12	[$4^2 6^4$]-b	bog	$4^2 6^4$-b	*mmm*	13
bph	14	[$4^6 6^3$]	afo	$4^6 6^3$	$\bar{6}2m$	3
bre	10	[$4^2 5^4$]	bru	$4^2 5^4$	*mmm*	9
can	18	[$4^6 6^5$]	can	$4^6 6^3 6^2$	*mm*2	18
cas	12	[$5^4 6^2$]	eun	$5^4 6^2$	2/*m*	20
cha	36	{3 [$4^{12}6^2 8^6$]}	cha	$4^6 4^6 6^2 8^6$	$\bar{3}2$	2
clo	32	{1 [$4^8 6^8 8^2$]}	rpa	$4^8 6^8 8^2$	$\bar{8}2m$	2
d4r	8	[4^6]	cub	4^6	4/$m\bar{3}m$	17
d6r	12	[$4^6 6^2$]	hpr	$4^6 6^2$	6/*mmm*	25
d8r	16	-	opr	$4^8 8^2$	8/*mmm*	5
gis	20	{2 [$4^6 8^4$]}	gsm	$4^4 4^2 8^4$-a	$\bar{4}2m$	3
gme	24	{2 [$4^9 6^2 8^3$]}	gme	$4^6 4^3 6^2 8^3$	$\bar{6}2m$	7
jbw	10	[6^4]	hes	6^4	$\bar{4}3m$	7
lau	12	[$4^2 6^4$]-a	lau	$4^2 6^4$-a	4/*mmm*	17
lio	42	[$4^6 6^{17}$]	lio	$4^6 6^6 6^6 6^3 6^2$	$\bar{6}2m$	5
los	30	[$4^6 6^{11}$]	los	$4^6 6^6 6^3 6^2$	$\bar{6}2m$	4
lov	5	-	sfi	3^2	$\bar{4}2m$	7
lta	48	{3 [$4^{12}6^8 8^6$]}	grc	$4^{12}6^8 8^6$	4/$m\bar{3}m$	8
ltl	48	{3 [$4^{18}8^6 12^2$]}	lil	$4^{12}4^6 8^6 12^2$-b	6/*mmm*	2
mei	7	-	iet	$4^3 6^1$	3*m*1	4
mel	14	[$4^1 5^2 6^4$]	mel	$4^1 5^2 6^2$	*mm*2	9
mfi	14	[5^8]	pen	$5^4 5^4$	$\bar{4}2m$	3
mor	8	[5^4]	tes	5^4-a	$\bar{4}2m$	14
mso	18	[6^8]-b	ber	$6^2 6^2$-a	6/*mmm*	2
mtn	20	[5^{12}]	red	5^{12}	53*m*	4
mtw	14	[$4^2 5^4 6^2$]	mtw	$4^2 5^4 6^2$	*mmm*	8
nat	6	[4^4]	des	4^4	$\bar{4}2m$	3
non	15	[$4^1 5^8$]	non	$4^1 5^4 5^2 5^2$	*mm*2	6
pau	32	{3 [$4^{12}8^6$]}	pau	$4^8 4^4 8^4 8^2$	4/*mmm*	4
rte	16	[$4^4 5^4 6^2$]	tte	$4^4 5^4 6^2$	*mmm*	2
rth	12	[$4^4 5^4$]	cle	$4^4 5^4$	$\bar{4}2m$	3
sod	24	[$4^6 6^8$]	toc	$4^6 6^8$	4/$m\bar{3}m$	10
stf	12	[$4^1 5^6$]	nuh	$4^1 5^2 5^2 5^2$	*mm*2	4
sti	8	[$4^6 6^1$]	sti	$4^2 4^2 6^1$	*mm*2	10

no equivalents for *bik* (13T, 7 occurences), *fer* (13T 5 occurences), *mtt* (14T, 9 occurrences), *ton* (15T, 9 occurences) or *vsv* (6T, 8 occurences)

[a] H. van Koningsveld, *Compendium of Zeolite Framework Types. Building Schemes and Type Characteristics*, Elsevier, Amsterdam, 2007
[b] J.V. Smith, *Tetrahedral frameworks of zeolites, clathrates and related materials*, Landolt-Börnstein, Vol. 14A, Springer, Berlin, 2000

APPENDIX E
Channel System Dimensions
ordered by largest ring

20-, 18- &14-Ring Structures

-CLO	Cloverite	<100> **20** 4.0 x 13.2*** | <100> **8** 3.8 x 3.8***
ETR	ECR-34	[001] **18** 10.1* ⊥ ⊥ [001] **8** 2.5 x6.0**
VFI	VPI-5	[001] **18** 12.7 x 12.7*
AET	AlPO-8	[001] **14** 7.9 x 8.7*
CFI	CIT-5	[010] **14** 7.2 x 7.5*
DON	UTD-1F	[010] **14** 8.1 x 8.2*
OSO	OSB-1	[001] **14** 5.4 x 7.3* ⊥ ⊥ [001] **8** 2.8 x 3.3**
SFH	SSZ-53	[001] **14** 6.4 x 8.7*
SFN	SSZ-59	[001] **14** 6.2 x 8.5*
UTL	IM-12	[001] **14** 7.1 x 9.5* ⊥ [010] **12** 5.5 x 8.5*

12-Ring Structures

AFI	AlPO-5	[001] **12** 7.3 x 7.3*
AFR	SAPO-40	[001] **12** 6.7 x 6.9* ⊥ [010] **8** 3.7 x 3.7*
AFS	MAPSO-46	[001] **12** 7.0 x 7.0* ⊥ ⊥ [001] **8** 4.0 x 4.0**
AFY	CoAPO-50	[001] **12** 6.1 x 6.1* ⊥ ⊥ [001] **8** 4.0 x 4.3**
ASV	ASU-7	[001] **12** 4.1x 4.1*
ATO	AlPO-31	[001] **12** 5.4 x 5.4*
ATS	MAPO-36	[001] **12** 6.5 x 7.5*
*BEA	Beta polymorph A	<100> **12** 6.6 x 6.7** ⊥ [001] **12** 5.6 x 5.6*
BEC	FOS-5 (Beta pol. C)	[001] **12** 6.3 x 7.5* ⊥ <100> **12** 6.0 x 6.9**
BOG	Boggsite	[100] **12** 7.0 x 7.0* ⊥ [010] **10** 5.5 x 5.8*
BPH	Beryllophosphate-H	[001] **12** 6.3 x 6.3* ⊥ ⊥ [001] **8** 2.7 x 3.5**
CAN	Cancrinite	[001] **12** 5.9 x 5.9*
CON	CIT-1	[001] **12** 6.4 x 7.0* ⊥ [100] **12** 7.0 x 5.9* ⊥ [010] **10** 5.1 x 4.5*
CZP	Chiral ZnPO$_4$	[001] **12** 3.8 x 7.2* (highly distorted 12-ring)
DFO	DAF-1	{[001] **12** 7.3 x 7.3 ⊥ ⊥ [001] **8** 3.4 x 5.6}*** ⊥ {[001] **12** 6.2 x 6.2 ⊥ ⊥ [001] **10** 5.4 x 6.4}***
EMT	EMC-2	[001] **12** 7.3 x 7.3* ⊥ ⊥ [001] **12** 6.5 x 7.5**

12-Ring Structures (cont.)

EON	ECR-1	{[100] **12** 6.7 x 6.8* ⊥ [010] **8** {[001] 3.4 x 4.9 ⊥ [100] **8** 2.9 x 2.9}*}**
EZT	EMM-3	[100] **12** 6.5 x 7.4*
FAU	Faujasite	<111> **12** 7.4 x 7.4***
GME	Gmelinite	[001] **12** 7.0 x 7.0* ⊥ ⊥ [001] **8** 3.6 x 3.9**
GON	GUS-1	[001] **12** 5.4 x 6.8*
IFR	ITQ-4	[001] **12** 6.2 x 7.2*
ISV	ITQ-7	<100> **12** 6.1 x 6.5** ⊥ [001] **12** 5.9 x 6.6*
IWR	ITQ-24	[001] **12** 5.8 x 6.8* ⊥ [110] **10** 4.6 x 5.3* ⊥ [010] **10** 4.6 x 5.3*
IWV	ITQ-27	{[001] **12** 6.2 x 6.9 ⊥ [011] **12** 6.2 x 6.9}**
IWW	ITQ-22	[001] **12** 6.0 x 6.7* ⊥ n[001] **10** 4.9 x 4.9** ⊥ [001] **8** 3.3 x 4.6*
LTL	Linde Type L	[001] **12** 7.1 x 7.1*
MAZ	Mazzite	[001] **12** 7.4 x 7.4* | [001] **8** 3.1 x 3.1***
MEI	ZSM-18	[001] **12** 6.9 x 6.9* ⊥ ⊥ [001] **7** 3.2 x 3.5**
MOR	Mordenite	[001] **12** 6.5 x 7.0* ⊥ [001] **8** 2.6 x 5.7***
MOZ	ZSM-10	{[001] **12** 6.8 x 7.0 ⊥ n[001] **8** 3.8 x 4.8}*** | [001] **12** 6.8 x 6.8*
MSE	MCM-68	{[001] **12** 6.4 x 6.8 ⊥ [100] **10** 5.2 x 5.8 ⊥ [110] **10** 5.2 x 5.2}***
MTW	ZSM-12	[010] **12** 5.6 x 6.0*
NPO	Nitridophosphate-1	[100] **12** 3.3 x 4.4*
OFF	Offretite	[001] **12** 6.7 x 6.8* ⊥ ⊥ [001] **8** 3.6 x 4.9**
OSI	UiO-6	[001] **12** 5.2 x 6.0*
-RON	Roggianite	[001] **12** 4.3 x 4.3*
RWY	UCR-20	<111> **12** 6.9 x 6.9***
SAO	STA-1	<100> **12** 6.5 x 7.2** ⊥ [001] **12** 7.0 x 7.0*
SBE	UCSB-8Co	<100> **12** 7.2 x 7.4** ⊥ [001] **8** 4.0 x 4.0*
SBS	UCSB-6GaCo	[001] **12** 6.8 x 6.8* ⊥ ⊥ [001] **12** 6.9 x 7.0**
SBT	UCSB-10GaZn	[001] **12** 6.4 x 7.4 * ⊥ ⊥ [001] **12** 7.3 x 7.8**
SFE	SSZ-48	[010] **12** 5.4 x 7.6*
SFO	SSZ-51	[001] **12** 6.9 x 7.1* ⊥ [010] **8** 3.1 x 3.9*
SOS	SU-16	{[100] **12** 3.9 x 9.1 ⊥ [010] **8** 3.3 x 3.3}**
SSY	SSZ-60	[001] **12** 5.0 x 7.6*

12-Ring Structures (cont.)

USI	IM-6	[100] **12** 6.1 x 6.2 * ⊥ [001] **10** 3.9 x 6.4*
VET	VPI-8	[001] **12** 5.9 x 5.9*

10-Ring Structures

AEL	AlPO-11	[001] **10** 4.0 x 6.5*
AFO	AlPO-41	[001] **10** 4.3 x 7.0*
AHT	AlPO-H2	[001] **10** 3.3 x 6.8*
CGF	Co-Ga-Phosphate-5	{[100] **10** 2.5 x 9.2* + **8** 2.1 x 6.7*} ⊥ [001] **8** 2.4 x 4.8*
CGS	Co-Ga-Phosphate-6	{[001] **10** 3.5 x 8.1 ⊥ [100] **8** 2.5 x 4.6}***
DAC	Dachiardite	[010] **10** 3.4 x 5.3* ⊥ [001] **8** 3.7 x 4.8*
EUO	EU-1	[100] **10** 4.1 x 5.4* with large side pockets
FER	Ferrierite	[001] **10** 4.2 x 5.4* ⊥ [010] **8** 3.5 x 4.8*
HEU	Heulandite	{[001] **10** 3.1 x 7.5* + **8** 3.6 x 4.6*} ⊥ [100] **8** 2.8x 4.7*
IMF	IM-5	{[001] **10** 5.5 x 5.6 ⊥ [100] **10** 5.3 x 5.4}** ⊥ {[010] **10** 5.3 x 5.9} ⊥ {[001] **10** 4.8 x 5.4 ⊥ [100] **10** 5.1 x 5.3}**
ITH	ITQ-13	[001] **10** 4.8 x 5.3* ⊥ [010] **10** 4.8 x 5.1* ⊥ [100] **9** 4.0 x 4.8*
LAU	Laumontite	[100] **10** 4.0 x 5.3* (contracts upon dehydration)
MEL	ZSM-11	<100> **10** 5.3 x 5.4***
MFI	ZSM-5	{[100] **10** 5.1 x 5.5 ⊥ [010] **10** 5.3 x 5.6}***
MFS	ZSM-57	[100] **10** 5.1 x 5.4* ⊥ [010] **8** 3.3 x 4.8*
MTT	ZSM-23	[001] **10** 4.5 x 5.2*
MWW	MCM-22	⊥ [001] **10** 4.0 x 5.5** | ⊥ [001] **10** 4.1 x 5.1**
NES	NU-87	[100] **10** 4.8 x 5.7**
OBW	OSB-2	{<110> **10** 5.0 x 5.0** ⊥ ([001] **8** 3.4 x 3.4* + <101> **8** 2.8 x 4.0**) ⊥ <100> 3.3 x 3.4**}***
-PAR	Partheite	[001] **10** 3.5 x 6.9*
PON	IST-1	[100] **10** 5.0 x 5.3*
RRO	RUB-41	[100] **10** 4.0 x 6.5* ⊥ [001] **8** 2.7 x 5.0*
SFF	SSZ-44	[001] **10** 5.4 x 5.7*
SFG	SSZ-58	[001] **10** 5.2 x 5.7* ⊥ [100] **10** 4.8 x 5.7*
STF	SSZ-35	[001] **10** 5.4 x 5.7*
STI	Stilbite	[100] **10** 4.7 x 5.0* ⊥ [001] **8** 2.7 x 5.6*
SZR	SUZ-4	{[001] **10** 4.1 x 5.2 ⊥ [010] **8** 3.2 x 4.8 ⊥ [110] **8** 3.0 x 4.8}***

10-Ring Structures (cont.)

TER	Terranovaite	[100] **10** 5.0 x 5.0* ⊥ [001] **10** 4.1 x 7.0*
TON	Theta-1	[001] **10** 4.6 x 5.7*
TUN	TNU-9	{[010] **10** 5.6 x 5.5 ⊥ [10$\overline{1}$] **10** 5.4 x 5.5}***
WEI	Weinebeneite	[001] **10** 3.1 x 5.4* ⊥ [100] **8** 3.3 x 5.0*
-WEN	Wenkite	<100> **10** 2.5 x 4.8** ⊥ [001] **8** 2.3 x 2.7*

9-Ring Structures

-CHI	Chiavennite	[001] **9** 3.9 x 4.3*
LOV	Lovdarite	[010] **9** 3.2 x 4.5* ⊥ [001] **9** 3.0 x 4.2* ⊥ [100] **8** 3.6 x 3.7*
NAB	Nabesite	[110] **9** 2.7 x 4.1* ⊥ [$\overline{1}\overline{1}$0] **9** 3.0 x 4.6*
NAT	Natrolite	<100> **8** 2.6 x 3.9** ⊥ [001] **9** 2.5 x 4.1*
RSN	RUB-17	[100] **9** 3.3 x 4.4* ⊥ [001] **9** 3.1 x 4.3* ⊥ [010] **8** 3.4 x 4.1*
STT	SSZ-23	[101] **9** 3.7 x 5.3* ⊥ [001] **7** 2.4 x 3.5*
VSV	VPI-7	[01-1] **9** 3.3 x 4.3* ⊥ [011] **9** 2.9 x 4.2* ⊥ [011] **8** 2.1 x 2.7*

8-Ring Structures

ABW	Li-A	[001] **8** 3.4 x 3.8*
ACO	ACP-1	<100> **8** 2.8 x 3.5** ⊥ [001] **8** 3.5 x 3.5*
AEI	AlPO-18	{[100] **8** 3.8 x 3.8 ⊥ [110] **8** 3.8 x 3.8 ⊥ [001] **8** 3.8 x 3.8}***
AEN	AlPO-EN3	[100] **8** 3.1 x 4.3* ⊥ [010] **8** 2.7 x 5.0*
AFN	AlPO-14	[100] **8** 1.9 x 4.6* ⊥ [010] **8** 2.1 x 4.9* ⊥ [001] **8** 3.3 x 4.0*
AFT	AlPO-52	⊥ [001] **8** 3.2 x 3.8***
AFX	SAPO-56	⊥ [001] **8** 3.4 x 3.6***
ANA	Analcime	irregular channels formed by highly distorted 8-rings
APC	AlPO-C	[001] **8** 3.4 x 3.7* ⊥ [100] **8** 2.0 x 4.7*
APD	AlPO-D	[010] **8** 2.3 x 6.0* ⊥ [201] **8** 1.3 x 5.8*
ATN	MAPO-39	[001] **8** 4.0 x 4.0*
ATT	AlPO-12-TAMU	[100] **8** 4.2 x 4.6* ⊥ [010] **8** 3.8 x 3.8*
ATV	AlPO-25	[001] **8** 3.0 x 4.9*
AWO	AlPO-21	[100] **8** 2.7 x 5.5*
AWW	AlPO-22	[001] **8** 3.9 x 3.9*
BCT	Mg-BCTT	[001] **8** 2.4 x 2.4*

8-Ring Structures (cont.)

BIK Bikitaite [010] **8** 2.8 x 3.7*

BRE Brewsterite [100] **8** 2.3 x 5.0* ⊥ [001] **8** 2.8 x 4.1*

CAS Cs Aluminosilicate [001] **8** 2.4 x 4.7*

CDO CDS-1 [010] **8** 3.1 x 4.7* ⊥ [001] **8** 2.5 x 4.2*

CHA Chabazite ⊥ [001] **8** 3.8 x 3.8***

DDR Deca-dodecasil 3R ⊥ [001] **8** 3.6 x 4.4**

DFT DAF-2 [001] **8** 4.1 x 4.1* ⊥ [100] **8** 1.8 x 4.7* ⊥ [010] **8** 1.8 x 4.7*

EAB TMA-E ⊥ [001] **8** 3.7 x 5.1**

EDI Edingtonite <110> **8** 2.8 x 3.8** ⊥ [001] **8** 2.0 x 3.1*

EPI Epistilbite {[001] **8** 3.7 x 4.5 ⊥ [100] **8** 3.6 x 3.6}**

ERI Erionite ⊥ [001] **8** 3.6 x 5.1***

ESV ERS-7 [010] **8** 3.5 x 4.7*

GIS Gismondine {[100] **8** 3.1 x 4.5 ⊥ [010] **8** 2.8 x 4.8}***

GOO Goosecreekite [100] **8** 2.8 x 4.0* ⊥ [010] **8** 2.7 x 4.1* ⊥ [001] **8** 2.9 x 4.7*

IHW ITQ-32 [100] **8** 3.5 x 4.3**

ITE ITQ-3 [010] **8** 3.8 x 4.3* ⊥ [001] **8** 2.7 x 5.8*

ITW ITQ-12 [100] **8** 2.4 x 5.4* ⊥ [001] **8** 3.9 x 4.2*

JBW Na-J [001] **8** 3.7 x 4.8*

KFI ZK-5 <100> **8** 3.9 x 3.9*** | <100> **8** 3.9 x 3.9***

LEV Levyne ⊥ [001] **8** 3.6 x 4.8**

LTA Linde Type A <100> **8** 4.1 x 4.1***

MER Merlinoite [100] **8** 3.1 x 3.5* ⊥ [010] **8** 2.7 x 3.6* ⊥
[001] {**8** 3.4 x 5.1* + **8** 3.3 x 3.3*}

MON Montesommaite [100] **8** 3.2 x 4.4* ⊥ [001] **8** 3.6 x 3.6*

MTF MCM-35 [001] **8** 3.6 x 3.9*

NSI Nu-6(2) [010] **8** 2.6 x 4.5* | [010] **8** 2.4 x 4.8*

OWE UiO-28 [010] **8** 3.5 x 4.0* ⊥ [001] **8** 3.2 x 4.8*

PAU Paulingite <100> **8** 3.6 x 3.6*** | <100> **8** 3.6 x 3.6***

PHI Phillipsite [100] **8** 3.8 x 3.8* ⊥ [010] **8** 3.0 x 4.3* ⊥ [001] **8** 3.2 x 3.3*

RHO Rho <100> **8** 3.6 x 3.6*** | <100> **8** 3.6 x 3.6***

RTE RUB-3 [001] **8** 3.7 x 4.4*

RTH RUB-13 [100] **8** 3.8 x 4.1* ⊥ [001] **8** 2.5 x 5.6*

8-Ring Structures (cont.)

RWR	RUB-24	[100] **8** 2.8 x 5.0* | [010] **8** 2.8 x 5.0*
SAS	STA-6	[001] **8** 4.2 x 4.2*
SAT	STA-2	⊥ [001] **8** 3.0 x 5.5***
SAV	Mg-STA-7	<100> **8** 3.8 x 3.8** ⊥ [001] **8** 3.9 x 3.9*
SIV	SIZ-7	{[100] **8** (3.5 x 3.9 + **8** 3.7 x 3.8 ⊥ [110] **8** 3.7 x 3.8 ⊥ [001] **8** 3.8 x 3.9} ***
THO	Thomsonite	[100] **8** 2.3 x 3.9* ⊥ [010] **8** 2.2 x 4.0* ⊥ [001] **8** 2.2 x 3.0*
TSC	Tschörtnerite	<100> **8** 4.2 x 4.2*** ⊥ <110> **8** 3.1 x 5.6***
UEI	Mu-18	{[010] **8** 3.5 x 4.6 ⊥ [001] **8** 2.5 x 3.6}**
UFI	UZM-5	<100> **8** 3.6 x 4.4** ⊥ [001] **8** 3.2 x 3.2(cage) i.e. ends in a cage
VNI	VPI-9	{<110> **8** 3.1 x 4.0 ⊥ [001] **8** 3.5 x .3.6}***
YUG	Yugawaralite	[100] **8** 2.8 x 3.6* ⊥ [001] **8** 3.1 x 5.0*
ZON	ZAPO-M1	[100] **8** 2.5 x 5.1* ⊥ [010] **8** 3.7 x 4.4*

APPENDIX F

Topological Densities

Code	$TD_{10}{}^a$	TD^b	Code	$TD_{10}{}^a$	TD^b	Code	$TD_{10}{}^a$	TD^b
ABW	833	0.703	-CHI	913	0.833	LAU	782	0.658
ACO	787	0.666	-CLO	456	0.443	LEV	719	0.605
AEI	689	0.583	CON	784	0.670	LIO	816	0.693
AEL	904	0.766	CZP	885	0.800	-LIT	768	0.655
AEN	956	0.857	DAC	977	0.841	LOS	816	0.693
AET	824	0.697	DDR	968	0.850	LOV	879	0.754
AFG	816	0.693	DFO	664	0.576	LTA	641	0.533
AFI	828	0.700	DFT	840	0.711	LTL	746	0.619
AFN	777	0.661	DOH	1002	0.882	LTN	779	0.698
AFO	907	0.769	DON	851	0.728	MAR	808	0.693
AFR	687	0.579	EAB	735	0.628	MAZ	823	0.697
AFS	656	0.568	EDI	786	0.666	MEI	728	0.630
AFT	685	0.585	EMT	584	0.493	MEL	944	0.808
AFX	689	0.585	EON	873	0.769	MEP	1059	0.955
AFY	585	0.488	EPI	979	0.845	MER	738	0.622
AHT	853	0.729	ERI	738	0.628	MFI	960	0.825
ANA	933	0.800	ESV	875	0.754	MFS	995	0.866
APC	814	0.696	ETR	675	0.576	MON	1033	0.885
APD	888	0.759	EUO	965	0.872	MOR	938	0.802
AST	742	0.625	EZT	754	0.639	MOZ	757	0.642
ASV	787	0.666	FAR	806	0.693	MSE	867	0.753
ATN	833	0.703	FAU	579	0.476	MSO	822	0.694
ATO	894	0.760	FER	1021	0.887	MTF	1083	0.942
ATS	752	0.640	FRA	802	0.683	MTN	1049	0.927
ATT	768	0.647	GIS	726	0.611	MTT	1015	0.883
ATV	960	0.816	GIU	811	0.693	MTW	912	0.776
AWO	828	0.708	GME	694	0.585	MWW	851	0.752
AWW	772	0.656	GON	926	0.787	NAB	873	0.740
BCT	959	0.814	GOO	840	0.716	NAT	834	0.740
*BEA	805	0.704	HEU	909	0.778	NES	922	0.818
BEC	764	0.656	IFR	798	0.678	NON	1038	0.915
BIK	1052	0.907	IHW	1013	0.898	NPO	869	0.750
BOG	781	0.659	IMF	964	0.846	NSI	1016	0.869
BPH	667	0.568	ISV	772	0.694	OBW	756	0.689
BRE	901	0.778	ITE	824	0.711	OFF	739	0.628
CAN	817	0.693	ITH	910	0.799	OSI	892	0.777
CAS	1042	0.895	ITW	901	0.769	OSO	747	0.645
CDO	1053	0.930	IWR	759	0.651	OWE	809	0.685
CFI	892	0.765	IWV	802	0.700	-PAR	773	0.664
CGF	819	0.695	IWW	849	0.756	PAU	728	0.623
CGS	718	0.613	JBW	890	0.753	PHI	751	0.635
CHA	677	0.566	KFI	681	0.571	PON	845	0.725

388

Code	TD_{10}^{a}	TD^{b}	Code	TD_{10}^{a}	TD^{b}	Code	TD_{10}^{a}	TD^{b}
RHO	641	0.533	SFF	880	0.765	TON	1006	0.867
-RON	771	0.688	SFG	932	0.850	TSC	590	0.482
RRO	924	0.801	SFH	835	0.724	TUN	929	0.803
RSN	914	0.786	SFN	819	0.692	UEI	832	0.708
RTE	844	0.715	SFO	681	0.575	UFI	747	0.641
RTH	817	0.695	SGT	962	0.862	UOZ	776	0.658
RUT	902	0.767	SIV	741	0.636	USI	691	0.590
RWR	1052	0.919	SOD	791	0.666	UTL	826	0.713
RWY	410	0.333	SOS	728	0.631	VET	1023	0.913
SAO	632	0.545	SSY	900	0.784	VFI	669	0.562
SAS	701	0.586	STF	877	0.748	VNI	971	0.896
SAT	763	0.644	STI	852	0.720	VSV	948	0.818
SAV	690	0.587	STT	859	0.760	WEI	773	0.655
SBE	619	0.514	SZR	910	0.780	-WEN	755	0.640
SBS	617	0.534	TER	872	0.739	YUG	935	0.797
SBT	617	0.522	THO	784	0.666	ZON	798	0.679
SFE	892	0.767	TOL	816	0.693			

[a] sum of all values of the coordination sequence from N_0 to N_{10}.
[b] topological density TD1000:100 as defined in the Explanatory Notes.

APPENDIX G

Origin of 3-Letter Codes and Material Names

Code	Abbreviated Name	Full Name
ABW		Li-A (Barrer and White)
ACO	ACP-1 (one)	Aluminium Cobalt Phosphate - one
AEI	$AlPO_4$-18 (eighteen)	Aluminophosphate-eighteen
AEL	$AlPO_4$-11 (eleven)	Aluminophosphate-eleven
AEN	AlPO-EN3	Aluminophosphate ethylenediamine (en) - three
AET	$AlPO_4$-8 (eight)	Aluminophosphate-eight
AFI	$AlPO_4$-5 (five)	Aluminophosphate-five
AFN	$AlPO_4$-14 (forteen)	Aluminophosphate-forteen
AFO	$AlPO_4$-41 (forty-one)	Aluminophosphate-forty-one
AFR	SAPO-40 (forty)	Silico-Aluminophosphate-forty
AFS	MAPSO-46 (forty-six)	$MgAl(P,Si)O_4$-46
AFT	$AlPO_4$-52 (fifty-two)	
AFX	SAPO-56 (fifty-six)	Silico-Aluminophosphate-fifty-six
AFY	CoAPO-50 (fifty)	
AHT	$AlPO_4$-H2 (two)	
APC	$AlPO_4$-C	
APD	$AlPO_4$-D	
AST	$AlPO_4$-16 (sixteen)	
ASV	ASU-7 (seven)	Arizona State University - seven
ATN	MAPO-39 (thirty-nine)	$MgAlPO_4$- thirty-nine
ATO	$AlPO_4$-31 (thirty-one)	
ATS	MAPO-36 (thirty-six)	
ATT	$AlPO_4$-12 (twelve)-TAMU	$AlPO_4$-12-Texas A&M University
ATV	$AlPO_4$-25 (twenty-five)	
AWO	$AlPO_4$-21 (twenty-one)	
AWW	$AlPO_4$-22 (twenty-two)	
*BEA	Zeolite Beta	
BPH	Beryllophosphate-H	Beryllophosphate-Harvey (or hexagonal)
CAS	Cesium Aluminsilicate	
CDO	CDS-1 (one)	Cylindrically Double Saw-edged structure - one
CFI	CIT-5 (five)	California Institute of Technology - five
CGF	CoGaPO-5 (five)	Cobalt-Gallium-Phosphate-five
CGS	CoGaPO-6 (six)	Cobalt-Gallium-Phosphate-six
-CLO	Cloverite	Four-leafed clover shaped pore opening
CON	CIT-1 (one)	California Institute of Technology - one
CZP		Chiral Zincophosphate
DDR	Deca-dodecasil 3R	Deca- & dodecahedra, 3 layers, rhombohedral
DFO	DAF-1 (one)	Davy Faraday Research Laboratory - one
DFT	DAF-2 (two)	Davy Faraday Research Laboratory - two
DOH	Dodecasil 1H	Dodecahedra, 1 layer, hexagonally stacked
DON	UTD-1 (one)	University of Texas at Dallas-one
EAB		TMA-E (Aiello and Barrer)
EMT	EMC-2 (two)	Elf (or Ecole Supérieure) Mulhouse Chimie - two

EON	ECR-1 (one)	Exxon Cooperate Research - one
ESV	ERS-7 (seven)	Eniricerche-molecular-sieve-seven
ETR	ECR-34 (thirty-four)	Exxon Cooperate Research - thirty-four
EUO	EU-1 (one)	Edinburgh Univerisity - one
EZT	EMM-3 (three) (Z for zeolite)	Exxon-Mobil Material - three
GIU	Giuseppettite	
GON	GUS-1 (one)	Gifu University Molecular Sieve - one
IFR	ITQ-4 (four)	Instituto de Tecnologia Quimica Valencia - four
IMF	IM-5 (five)	Institut Français du Pétrole and University of Mulhouse - five
ISV	ITQ-7 (seven)	Instituto de Tecnologia Quimica Valencia - seven
ITE	ITQ-3 (three)	Instituto de Tecnologia Quimica Valencia - three
ITH	ITQ-13 (thirteen)	Instituto de Tecnologia Quimica Valencia - thirteen
ITW	ITQ-12 (twelve)	Instituto de Tecnologia Quimica Valencia - twelve
IWR	ITQ-24 (twenty-four)	Instituto de Tecnologia Quimica Valencia - twenty-four
IWV	ITQ-27 (twenty-seven)	Instituto de Tecnologia Quimica Valencia - twenty-seven
IWW	ITQ-22 (twenty-two)	Instituto de Tecnologia Quimica Valencia - twenty-two
JBW		Na-J (Barrer and White)
KFI	ZK-5 (five)	Zeolite Kerr - five
LOS	Losod	Low sodium aluminosilicate
LTA	Linde Type A	Zeolite A (Linde Division, Union Carbide)
LTL	Linde Type L	Zeolite L (Linde Division, Union Carbide)
LTN	Linde Type N	Zeolite N (Linde Division, Union Carbide)
MEI	ZSM-18 (eighteen)	Zeolite Socony Mobil - eighteen
MEL	ZSM-11 (eleven)	Zeolite Socony Mobil - eleven
MFI	ZSM-5 (five)	Zeolite Socony Mobil - five
MFS	ZSM-57 (fifty-seven)	Zeolite Socony Mobil - fifty-seven
MSE	MCM-68 (sixty-eight)	Mobil Composition of Matter - sixty-eight
MSO	MCM-61 (sixty-one)	Mobil Composition of Matter - sixty-one
MTF	MCM-35 (thirty-five)	Mobil Composition of Matter - thirty-five
MTN	ZSM-39 (thirty-nine)	Zeolite Socony Mobil - thirty-nine
MTT	ZSM-23 (twenty-three)	Zeolite Socony Mobil - twenty-three
MTW	ZSM-12 (twelve)	Zeolite Socony Mobil - twelve
MWW	MCM-22 (twenty-two)	Mobil Composition of Matter-twenty-two
NES	NU-87 (eighty-seven)	New (ICI) - eighty-seven
NON	Nonasil	Nonahedra, all silica composition
NPO	Oxonitridophosphate-1 (one)	
NSI	Nu-6(2) (six)	New (ICI) - six
OBW	OSB-2 (two)	Universities of Oslo and Calif., Santa Barbara - two
OSI	UiO-6 (six)	University of Oslo - six
OSO	OSB-1 (one)	Universities of Oslo and Calif., Santa Barbara - one
PON	IST-1, Portugal Number One	Instituto Superior Técnico, Lisboa, - one
-RON	Roggianite	
RRO	RUB-41 (forty-one)	Ruhr University Bochum - forty-one
RSN	RUB-17 (seventeen)	Ruhr University Bochum - seventeen
RTE	RUB-3 (three)	Ruhr University Bochum - three
RTH	RUB-13 (thirteen)	Ruhr University Bochum - thirteen
RUT	RUB-10 (ten)	Ruhr University Bochum - ten

RWR	RUB-24 (twenty-four)	Ruhr University Bochum - twenty-four
RWY	UCR-20 (twenty)	University of California, Riverside -twenty
SAO	STA-1 (one)	University of Saint Andrews - one
SAS	STA-6 (six)	University of Saint Andrews - six
SAT	STA-2 (two)	University of Saint Andrews - two
SAV	Mg-STA-7 (seven)	University of Saint Andrews - seven
SBE	UCSB-8 (eight)	University of California, Santa Barbara - eight
SBS	UCSB-6 (six)	University of California, Santa Barbara - six
SBT	UCSB-10 (ten)	University of California, Santa Barbara - ten
SFE	SSZ-48 (forty-eight)	Standard Oil Synthetic Zeolite - fourty-eight
SFF	SSZ-44 (forty-four)	Standard Oil Synthetic Zeolite - fourty-four
SFG	SSZ-58 (fifty-eight)	Standard Oil Synthetic Zeolite - fifty-eight
SFH	SSZ-53 (fifty-three)	Standard Oil Synthetic Zeolite - fifty-three
SFN	SSZ-59 (fifty-nine)	Standard Oil Synthetic Zeolite - fifty-nine
SFO	SSZ-51 (fifty-one)	Standard Oil Synthetic Zeolite - fifty-one
SGT	Sigma-2 (two)	
SIV	SIZ-7 (seven)	St. Andrews Ionothermal Zeolite - seven
SOS	SU-16 (one-six)	Stockholm University - sixteen
STF	SSZ-35 (thirty-five)	Standard Oil Synthetic Zeolite - thirty-five
STT	SSZ-23 (twenty-three)	Standard Oil Synthetic Zeolite - twenty-three
SSY	SSZ-60 (sixty)	Standard Oil Synthetic Zeolite - sixty
SZR	SUZ-4 (four)	Sunbury Zeolite - four
TOL	Tounkite-like mineral	
TON	Theta-1 (one)	
TUN	TNU-9 (nine)	Taejon National University - nine
UEI	MU-18 (eighteen)	Mulhouse - eighteen
UFI	UZM-5 (five)	UOP Zeolitic Material - five
UOZ	IM-10, Mulhouse (one-zero)	Institut Français du Pétrole and University of Mulhouse - ten
USI	IM-6, Mulhouse (six)	Institut Français du Pétrole and University of Mulhouse - six
UTL	IM-12, Mulhouse (twelve)	Institut Français du Pétrole and University of Mulhouse - twelve
UWE	UiO-28 (twenty-eight)	University of Oslo - twenty-eight
VET	VPI-8 (eight)	Virgina Polytechnic Institute - eight
VFI	VPI-5 (five)	Virgina Polytechnic Institute - five
VNI	VPI-9 (nine)	Virgina Polytechnic Institute - nine
VSV	VPI-7 (seven)	Virgina Polytechnic Institute - seven
ZON	ZAPO-M1 (one)	$(Zn,Al)PO_4$ -Mulhouse - one

Isotypic Material Index

*ACP-1	ACO	*AlPO-H2	AHT
ACP-2	OWE	*AlPO-H2	AHT
ACP-3	DFT	AlPO-H3	APC
*Afghanite	AFG	Amicite	GIS
Al-ITQ-13	ITH	Ammonioleucite	ANA
Al-Nu-1	RUT	AMS-1B	MFI
Al-rich beta	*BEA	*Analcime	ANA
Alpha	LTA	*Analcime	ANA
AlPO-pollucite	ANA	*Analcime	ANA
*AlPO-11	AEL	APO-CJ3	APD
*AlPO-12-TAMU	ATT	*ASU-7	ASV
*AlPO-14	AFN	AZ-1	MFI
*AlPO-16	AST	B-Nu-1	RUT
AlPO-17	ERI	Barrerite	STI
*AlPO-18	AEI	Basic cancrinite	CAN
AlPO-20	SOD	Basic sodalite	SOD
*AlPO-21	AWO	Bellbergite	EAB
*AlPO-21	AWO	Beryllophosphate X	FAU
*AlPO-22	AWW	*Beryllophosphate-H	BPH
AlPO-24	ANA	*Beta	*BEA
*AlPO-25	ATV	Bicchulite	SOD
*AlPO-31	ATO	*Bikitaite	BIK
*AlPO-31	ATO	*Boggsite	BOG
AlPO-33	ATT	Bor-C	MFI
AlPO-33	ATT	Bor-D	MEL/MFI
AlPO-34	CHA	Boralite C	MFI
AlPO-35	LEV	Boralite D	MEL
AlPO-36	ATS	*Brewsterite	BRE
AlPO-40	AFR	Bystrite	LOS
AlPO-40	AFR	Ca-D	ANA
*AlPO-41	AFO	Ca-Q	MOR
*AlPO-5	AFI	*Cancrinite	CAN
*AlPO-52	AFT	Cancrinite hydrate	CAN
*AlPO-52	AFT	*CDS-1	CDO
AlPO-53(A)	AEN	*Cesium Aluminosilicate	CAS
AlPO-53(B)	AEN	CF-3	NON
AlPO-54	VFI	CF-4	MTN
*AlPO-8	AET	CFSAPO-1A	AEN
*AlPO-8	AET	*Chabazite	CHA
*AlPO-C	APC	*Chiavennite	-CHI
*AlPO-C	APC	*Chiral Zincophosphate	CZP
AlPO-CJB1	AWW	*CIT-1	CON
*AlPO-D	APD	CIT-4	BRE
*AlPO-EN3	AEN	*CIT-5	CFI
*AlPO-H2	AHT	CIT-6	*BEA

394

Clinoptilolite	**HEU**	*Erionite	**ERI**
*Cloverite	**-CLO**	*ERS-7	**ESV**
*Co-Ga-Phosphate-5	**CGF**	*EU-1	**EUO**
*Co-Ga-Phosphate-6	**CGS**	EU-13	**MTT**
Co-STA-7	**SAV**	EU-20	**CAS/NSI**
CoAPO-40	**AFR**	EU-20b	**CAS/NSI**
CoAPO-44	**CHA**	FAPO-36	**ATS**
CoAPO-47	**CHA**	FAPO-5	**AFI**
CoAPO-5	**AFI**	FAPO-H1	**VFI**
*CoAPO-50	**AFY**	*Farneseite	**FAR**
CoAPO-H3	**APC**	*Faujasite	**FAU**
CoAPSO-40	**AFR**	Fe(III)-BCTT	**BCT**
CoDAF-4	**LEV**	Fe-Nu-1	**RUT**
CoIST-2	**AEN**	*Ferrierite	**FER**
COK-5	**MFS**	FJ-17	**SOS**
CoVPI-5	**VFI**	*FOS-5	**BEC**
CrAPO-5	**AFI**	*Franzinite	**FRA**
CSZ-1	**EMT/FAU**	FU-9	**FER**
CZH-5	**MTW**	FZ-1	**MFI**
*Dachiardite	**DAC**	G	**SOD**
*DAF-1	**DFO**	Ga-Nu-1	**RUT**
*DAF-2	**DFT**	Gallosilicate ECR-10	**RHO**
DAF-5	**CHA**	Gallosilicate L	**LTL**
DAF-8	**PHI**	GaPO-14	**AFN**
Danalite	**SOD**	GaPO-34	**CHA**
Davyne	**CAN**	GaPO-DAB-2	**ZON**
*Deca-dodecasil 3R	**DDR**	Garronite	**GIS**
Dehyd. boggsite	**BOG**	Gaultite	**VSV**
Dehyd. brewsterite	**BRE**	GeAPO-11	**AEL**
Dehyd. Ca,NH4-Heulandite	**HEU**	Genthelvite	**SOD**
Dehyd. Linde Type A	**LTA**	*Gismondine	**GIS**
Dehyd. Na-Chabazite	**CHA**	*Giuseppettite	**GIU**
Dehyd. Na-X	**FAU**	*Gmelinite	**GME**
Dehyd. US-Y	**FAU**	Gobbinsite	**GIS**
Deuterated Rho	**RHO**	Gonnardite	**NAT**
*Dodecasil 1H	**DOH**	*Goosecreekite	**GOO**
Dodecasil-3C	**MTN**	Gottardiite	**NES**
*ECR-1	**EON**	*GUS-1	**GON**
ECR-18	**PAU**	H1	**VFI**
ECR-30	**EMT/FAU**	Harmotome	**PHI**
*ECR-34	**ETR**	Hauyn	**SOD**
ECR-40	**MEI**	Helvin	**SOD**
ECR-5	**CAN**	*Heulandite	**HEU**
*Edingtonite	**EDI**	Heulandite-Ba	**HEU**
*EMC-2	**EMT**	High natrolite	**NAT**
*EMM-3	**EZT**	High-silica Na-P	**GIS**
Encilite	**MFI**	Holdstite	**MTN**
*Epistilbite	**EPI**	Hsianghualite	**ANA**
ERB-1	**MWW**	Hydroxo sodalite	**SOD**

*IM-10	**UOZ**	*Liottite	**LIO**
*IM-12	**UTL**	*Lithosite	**-LIT**
*IM-5	**IMF**	*Losod	**LOS**
*IM-6	**USI**	*Lovdarite	**LOV**
IM-7	**ITH**	low melanophlogite	**MEP**
ISI-1	**TON**	Low-silica Na-P (MAP)	**GIS**
ISI-4	**MTT**	LZ-105	**MFI**
ISI-6	**FER**	LZ-132	**LEV**
*IST-1	**PON**	LZ-202	**MAZ**
IST-2	**AEN**	LZ-210	**FAU**
ITQ-1	**MWW**	LZ-211	**MOR**
*ITQ-12	**ITW**	LZ-212	**LTL**
*ITQ-13	**ITH**	LZ-214	**RHO**
ITQ-14	**BEC**	LZ-215	**LTA**
ITQ-15	**UTL**	LZ-217	**OFF**
ITQ-17	**BEC**	LZ-218	**CHA**
*ITQ-22	**IWW**	LZ-219	**HEU**
*ITQ-24	**IWR**	LZ-220	**ERI**
*ITQ-27	**IWV**	MAP	**GIS**
ITQ-29	**LTA**	MAPO-31	**ATO**
*ITQ-3	**ITE**	*MAPO-36	**ATS**
*ITQ-32	**IHW**	*MAPO-39	**ATN**
*ITQ-4	**IFR**	MAPO-43	**GIS**
*ITQ-4	**IFR**	MAPO-46	**AFS**
*ITQ-7	**ISV**	MAPO-5	**AFI**
ITQ-9	**STF**	MAPSO-43	**GIS**
JDF-2	**AEN**	*MAPSO-46	**AFS**
(K,Ba)-G,L	**LTL**	MAPSO-56	**AFX**
K-F	**EDI**	Maricopaite	**MOR**
K-M	**MER**	*Marinellite	**MAR**
K-rich gmelinite	**GME**	*Mazzite	**MAZ**
KZ-1	**MTT**	*MCM-22	**MWW**
KZ-2	**TON**	*MCM-35	**MTF**
Large port mordenite	**MOR**	MCM-37	**AET**
*Laumontite	**LAU**	MCM-58	**IFR**
Leonhardite	**LAU**	*MCM-61	**MSO**
Leucite	**ANA**	MCM-65	**CDO**
*Levyne	**LEV**	*MCM-68	**MSE**
*Li-A	**ABW**	MCM-9	**VFI**
Li-LSX	**FAU**	MCS-1	**AEN**
Linde D	**CHA**	MeAPO-47	**CHA**
Linde F	**EDI**	MeAPSO-47	**CHA**
Linde Q	**BPH**	*Melanophlogite	**MEP**
Linde R	**CHA**	*Merlinoite	**MER**
Linde T	**ERI/OFF**	Mesolite	**NAT**
*Linde Type A (zeolite A)	**LTA**	Metanatrolite	**NAT**
*Linde Type L (zeolite L)	**LTL**	Metavariscite	**BCT**
*Linde Type N	**LTN**	*Mg-BCTT	**BCT**
Linde W	**MER**	*Mg-STA-7	**SAV**

MgAPO-50	**AFY**	*Paulingite	**PAU**
Microsommite	**CAN**	Perlialite	**LTL**
MnAPO-11	**AEL**	Phi	**CHA**
MnAPO-14	**AEN**	*Phillipsite	**PHI**
MnAPO-41	**AFO**	Pollucite	**ANA**
MnAPO-50	**AFY**	Primary leonhardite	**LAU**
MnAPSO-41	**AFO**	PSH-3	**MWW**
*Montesommaite	**MON**	Q	**KFI**
*Mordenite	**MOR**	*Rho	**RHO**
Mu-10	**AEN**	RMA-3	**ATT**
Mu-13	**MSO**	*Roggianite	**-RON**
Mu-14	**ITE**	RUB-1	**LEV**
*Mu-18	**UEI**	*RUB-10	**RUT**
Mu-26	**STF**	*RUB-13	**RTH**
Mutinaite	**MFI**	*RUB-17	**RSN**
N-A	**LTA**	*RUB-24	**RWR**
Na-B	**ANA**	*RUB-24	**RWR**
Na-D	**MOR**	*RUB-3	**RTE**
*Na-J	**JBW**	*RUB-3	**RTE**
Na-P1	**GIS**	*RUB-41	**RRO**
Na-P2	**GIS**	SAPO-11	**AEL**
Na-V	**THO**	SAPO-18	**AEI**
*Nabesite	**NAB**	SAPO-31	**ATO**
*Natrolite	**NAT**	SAPO-31	**ATO**
NaZ-21	**LTN**	SAPO-34	**CHA**
Nepheline hydrate	**JBW**	SAPO-35	**LEV**
(Ni(deta)2)-UT-6	**CHA**	SAPO-37	**FAU**
*Nitridophosphate-1	**NPO**	SAPO-39	**ATN**
*Nonasil	**NON**	*SAPO-40	**AFR**
Nosean	**SOD**	SAPO-41	**AFO**
Nu-1	**RUT**	SAPO-42	**LTA**
NU-10	**TON**	SAPO-43	**GIS**
NU-13	**MTW**	SAPO-47	**CHA**
NU-23	**FER**	SAPO-5	**AFI**
NU-3	**LEV**	*SAPO-56	**AFX**
NU-4	**MFI**	Scolecite	**NAT**
NU-5	**MFI**	Sigma-1	**DDR**
*Nu-6(2)	**NSI**	*Sigma-2	**SGT**
*NU-87	**NES**	Silicalite	**MFI**
o-FDBDM-ZSM-50	**EUO**	Silicalite 2	**MEL**
Octadecasil	**AST**	Siliceous Na-Y	**FAU**
*Offretite	**OFF**	*SIZ-7	**SIV**
Omega	**MAZ**	SIZ-8	**AEI**
*OSB-1	**OSO**	SIZ-9	**SOD**
*OSB-2	**OBW**	*Sodalite	**SOD**
P	**KFI**	Sr-D	**FER**
Pahasapaite	**RHO**	Sr-Q	**YUG**
Paranatrolite	**NAT**	SSZ-16	**AFX**
*Partheite	**-PAR**	*SSZ-23	**STT**

SSZ-24	**AFI**	Synthetic stellerite	**STI**
SSZ-25	**MWW**	Synthetic stilbite	**STI**
SSZ-26	**CON**	Synthetic thomsonite	**THO**
SSZ-33	**CON**	Synthetic wairakite	**ANA**
*SSZ-35	**STF**	*Terranovaite	**TER**
SSZ-36	**ITE/RTH**	Tetragonal edingtonite	**EDI**
SSZ-39	**AEI**	Tetranatrolite	**NAT**
SSZ-42	**IFR**	*Theta-1	**TON**
*SSZ-44	**SFF**	Theta-3	**MTW**
SSZ-46	**MEL**	*Thomsonite	**THO**
*SSZ-48	**SFE**	Tiptopite	**CAN**
SSZ-50	**RTH**	TiVPI-5	**VFI**
*SSZ-51	**SFO**	TMA sodalite	**SOD**
*SSZ-53	**SFH**	*TMA-E	**EAB**
SSZ-55	**ATS**	TMA-gismondine	**GIS**
*SSZ-58	**SFG**	TMA-O	**OFF**
*SSZ-59	**SFN**	TNU-1	**CGS**
*SSZ-60	**SSY**	TNU-10	**STI**
SSZ-73	**SAS**	TNU-7	**EON**
*STA-1	**SAO**	*TNU-9	**TUN**
*STA-2	**SAT**	*Tounkite-like mineral	**TOL**
STA-5	**BPH**	TPZ-12	**MTW**
*STA-6	**SAS**	TPZ-3	**EUO**
Stellerite	**STI**	Triclinic bikitaite	**BIK**
*Stilbite	**STI**	TS-1	**MFI**
*SU-16	**SOS**	TS-2	**MEL**
*SUZ-4	**SZR**	*Tschörtnerite	**TSC**
Svetlozarite	**DAC**	Tschernichite	***BEA**
Svyatoslavite	**BCT**	TsG-1	**CGS**
Synthetic amicite	**GIS**	TSZ	**MFI**
Synthetic analcime	**ANA**	TSZ-III	**MFI**
Synthetic barrerite	**STI**	Tugtupite	**SOD**
Synthetic brewsterite	**BRE**	TZ-01	**MFI**
Synthetic Ca-garronite	**GIS**	*UCR-20	**RWY**
Synthetic cancrinite	**CAN**	UCSB-10Co	**SBT**
Synthetic edingtonite	**EDI**	*UCSB-10GaZn	**SBT**
Synthetic epistilbite	**EPI**	UCSB-10Mg	**SBT**
Synthetic garronite	**GIS**	UCSB-10Zn	**SBT**
Synthetic gobbinsite	**GIS**	UCSB-3	**ABW**
Synthetic gonnardite	**NAT**	UCSB-3GaGe	**DFT**
Synthetic hsinghualite	**ANA**	UCSB-3ZnAs	**DFT**
Synthetic laumontite	**LAU**	UCSB-6Co	**SBS**
Synthetic lovdarite	**LOV**	*UCSB-6GaCo	**SBS**
Synthetic melanophlogite	**MEP**	UCSB-6GaMg	**SBS**
Synthetic merlinoite	**MER**	UCSB-6GaZn	**SBS**
Synthetic mesolite	**NAT**	UCSB-6Mg	**SBS**
Synthetic natrolite	**NAT**	UCSB-6Mn	**SBS**
Synthetic offretite	**OFF**	UCSB-6Zn	**SBS**
Synthetic scolecite	**NAT**	*UCSB-8Co	**SBE**

UCSB-8Mg	**SBE**	Zeolite X (Linde X)	**FAU**
UCSB-8Mn	**SBE**	Zeolite Y (Linde Y)	**FAU**
UCSB-8Zn	**SBE**	Zincophosphate X	**FAU**
UiO-12-500	**AEN**	ZK-14	**CHA**
UiO-12-as	**AEN**	ZK-19	**PHI**
UiO-20	**DFT**	ZK-20	**LEV**
UiO-21	**CHA**	ZK-21	**LTA**
*UiO-28	**OWE**	ZK-22	**LTA**
*UiO-6	**OSI**	ZK-4	**LTA**
UiO-7	**ZON**	*ZK-5	**KFI**
UiO-7	**ZON**	ZKQ-1B	**MFI**
USC-4	**MFI**	ZMQ-TB	**MFI**
USI-108	**MFI**	Zn-BCTT	**BCT**
UTD-1	**DON**	Zn-STA-7	**SAV**
*UTD-1F	**DON**	ZnAPO-35	**LEV**
UTM-1	**MTF**	ZnAPO-36	**ATS**
UZM-25	**CDO**	ZnAPO-39	**ATN**
UZM-4	**BPH**	ZnAPO-40	**AFR**
*UZM-5	**UFI**	ZnAPO-5	**AFI**
VAPO-31	**ATO**	ZnAPO-50	**AFY**
VAPO-5	**AFI**	ZnAPSO-40	**AFR**
Vishnevite	**CAN**	*ZSM-10	**MOZ**
*VPI-5	**VFI**	*ZSM-11	**MEL**
*VPI-7	**VSV**	*ZSM-12	**MTW**
*VPI-8	**VET**	*ZSM-18	**MEI**
*VPI-9	**VNI**	ZSM-20	**EMT/FAU**
VS-12	**MTW**	ZSM-22	**TON**
VSV-7#	**VSV**	*ZSM-23	**MTT**
Wairakite	**ANA**	ZSM-3	**EMT/FAU**
*Weinebeneite	**WEI**	ZSM-35	**FER**
Wellsite	**PHI**	*ZSM-39	**MTN**
*Wenkite	**-WEN**	ZSM-4	**MAZ**
Willhendersonite	**CHA**	*ZSM-5	**MFI**
*Yugawaralite	**YUG**	ZSM-50	**EUO**
*ZAPO-M1	**ZON**	ZSM-51	**NON**
ZBH	**MFI**	*ZSM-57	**MFS**
ZCP-THO	**THO**	ZSM-58	**DDR**
Zeolite N	**EDI**	ZYT-6	**CHA**
Zeolite W	**MER**		

Printed and bound by CPI Group (UK) Ltd, Croydon, CR0 4YY

03/10/2024

01040331-0019